PRINCIPLES OF SPINNING

PRINCIPLES OF SPINNING
Fibres and Blow Room Cotton Processing in Spinning

Ashok R. Khare

CRC Press
Taylor & Francis Group
Boca Raton London New York

CRC Press is an imprint of the
Taylor & Francis Group, an **informa** business

First edition published 2022 by
CRC Press
6000 Broken Sound Parkway NW, Suite 300, Boca Raton, FL 33487-2742

and by
CRC Press
2 Park Square, Milton Park, Abingdon, Oxon, OX14 4RN

ISBN: 978-1-138-59291-9 (hbk)
ISBN: 978-0-367-68878-3 (pbk)
ISBN: 978-0-429-48658-6 (ebk)

Typeset in Times
by MPS Limited, Dehradun

Contents

List of Tables

Preface

In my earlier venture 'Raw Cotton and Blow Room' I made sincere efforts to provide very useful and valuable basic information to the student community in India. Being a teacher, my intentions were to provide all possible information to give them a thorough knowledge of the process as in a classroom session based on their syllabus.

I received an overwhelming response from the students from different parts of the country. The students found the information explained in a very lucid way. They were able to understand almost all the technical data as if it was taught to them in the class. This, I feel, is the success of my efforts.

Over the years, technology has rapidly changed. In this book, while retaining some basics, I have tried to rehash the form to keep pace with time and changing technology. Thus, the chapter on modern blow room and related concept in implementing the modern techniques in processing cotton through blow room have been incorporated in this book. The machines in modern blow room sequence have adopted these research guidelines and, thus, the modern blow room line sequence has not only been automated but also much shortened. This, therefore, has substantially reduced the potential danger of over-beating and over-treating the cotton. The modern cleaning devices are now referred to as opening devices, which have the advantage of using this intensive opening for cleaning the cotton material in the same machines itself.

Even at the start of Blow Room, the most popular concept of giving pre-opening to hard-matted mass of cotton by some preliminary beaters has been totally revolutionized by the introduction of 'Automatic Bale Opening' machine (Blendomat and similar kinds). This has enabled replacement of initial few machines used in conventional blow room. The impetus that the cotton material receives after this process continues thereafter. This is possible only because of the great intensive opening that the cotton receives in automatic bale opener. In addition to this, the machines like CVT cleaners have made a remarkable improvement in tackling the problem of trash extraction, in addition to continued opening of cotton tufts.

Modern electronics has also entered into management of various controls and now the interface devices have made it possible to take the help of 'user friendly' on-line controls. With them, it is possible to control the machine action with precision. Thus, it is now possible, without stopping the machine, to set a pre-determined proportion of lint and trash percentage in blow room dropping. The chute feeding to card is yet another feature that has not only reduced an additional space required to store the blow room laps but has also helped in maintaining the same level of openness that cotton tufts reach at the end of the blow room sequence.

Finally, all this has helped in making the whole blow room line much compact, with very few machines needed to complete its job. It has also enabled to substantially reduce the labour complement in this whole department. In this book, I have again sincerely tried to bring out these different features in my own style, and I do hope that they will be immensely useful to the student community.

Dr. Ashok Khare,
Mumbai 400016

Acknowledgement

Since I was a student in college, I had always dreamt of becoming a teacher. I idolized some of my teachers then. I feel proud to mention their names. This is because, later when I became a teacher in the same college after several years, I always found inspiration when remembering them, their style and philosophy of their teaching, their sincerity and devotion to the profession and their skill in making the difficult things look simple. All of them had a good industrial experience and were, therefore, able to share their knowledge with the students.

Prof. D.B. Ajgaokar, who later became the first Principal of D.K.T.E. Institute Ichalkaranji, encouraged me to venture into writing books for students. After my first venture, he continued his encouragement and support for writing more. This is how I was able to take up this vast task of writing a book-series on spinning technology.

Late Prof. M.K. Naboodiri had been my philosopher and guide and very few would really know that I had family-like relation with him. When any problem was posed to him, he would never offer haphazard answers. He would meditate and then come out with a logical solution.

Dr. V.S. Jayram was my first teacher, when I started learning about textile technology, a field which was unheard till the completion of my school days. I immensely liked the way he taught spinning. I must admit that the credit of my continuing with textile studies solely goes to him. As a student, I never missed his lectures. I must also confess that his style of teaching inspired me to come back to my college as a teacher. When I requested him to edit this volume, he very gladly accepted and completed the work. But most unfortunately, he passed away after a short period. I will always remain indebted to Dr. Jayram.

Dr. S.G. Vinzanekar, my mentor during my whole career as a teacher, was always a source of inspiration for giving me several opportunities to learn things. When he would entrust any job, he always used to fully back and support me, irrespective of whether it was a success or otherwise and even in odd circumstances. He was my guide for my Ph.D. work and thereon, he became my senior philosopher friend.

I am extremely thankful to Director CIRCOT (Central Institute for Research on Cotton Technology), Mumbai, for helping me from time to time, to make the CIRCOT research work on cotton technology available. The thanks are very much due to two giant and reputed machinery manufacturers – M/s Trumac-Trutzschler & M/s Rieters for providing me with beautiful diagrams supporting the theory of blow room machines. Without their help and permission, this book would not have been real worth, which I feel, is what it is now.

The core of this book appears to be similar to Manual of Cotton Spinning book-series, published long back by Shirley Institute, though in treatment to the subject-information, it differs. I sincerely and profusely thank The Textile (formerly Shirley) Institute, Manchester, and London for giving me the inspiration to write and add some useful information to the ocean of textiles. Equally important was the permission from M/s Elsevier (former Butterworth publication), who permitted me to refer to their book 'Spun Yarn Technologies' (by Eric Oxtoby). I am greatly indebted to them. I hope, in the present form, it is still useful to students. The thanks are also due to my well-wishers, who directly or indirectly helped me in this venture.

Last but not the least, I would be failing in my duties if I do not mention the name of my father, late Shri Ramchandra Narayan Khare, who since my childhood groomed me to become a good student, good teacher and a good citizen. He himself was a born teacher and expert in child psychology. When I was thinking of leaving my job from the industry to join a teaching career, he asked me only one question – Would you take up this career with all sincerity and dedication? He also imbibed in my mind that if I have to become a good teacher, I always need to be a good student. I have never forgotten his words during my whole professional career as a teacher.

If I have authored this book, the credit goes to all these great personalities; as in some way or other, they have been instrumental in making me as I stand today.

Dr. Ashok R. Khare
Mumbai – 400 019

About the Author

Dr. Ashok Khare is a graduate, post-graduate and doctorate from a well-known technological institute – Veermata Jijabai Technological Institute, Mumbai (formerly known as Victoria Jubilee Technical Institute). He graduated from this Institute in 1970 and went to serve a well-known textile group of Mafatlal mills. After serving for nearly 5 years in the textile mills, he returned back to his Alma mater in 1975 as a lecturer in textile technology. In the due course, he was promoted to Assistant Professor and Professor.

In the last phase of his service in V.J.T.I., Mumbai, he took over as the Head of the Textile Manufacture's Department till he retired in 2006. Almost during the same tenure, he held the position of Deputy Director in the same Institute. He has written several articles on card cleaning efficiency, role of uni-comb, and extended research work on the influence of doubling parameters on properties of blended doubles yarns.

1 Fibres

1.1 HISTORY OF FIBRES

Hair-like material in a continuous filament form or in a discrete elongated form, similar in length to pieces of threads, is called 'fibres'. It can be converted into indiscrete filaments, threads or ropes. These can be used as material in composites, and also matted into a sheet to make papers or felts. There are basically of three types: natural fibres, fibres extracted from cellulose and purely synthetic fibres. The earliest evidence of humans using fibres goes with the use of wool and flax fibres which, sometimes in the dyed form, were located in prehistoric caves.

Since ancient times, therefore, the art of making yarns and fabric-like wares is known. In those days, China and Japan were specialists in making very fine silk fabrics. It was reported that some of the garments made of silk were so fine that the whole garment could easily pass through a small orifice of a wedding ring. After using fibres of animal origin, humans started using fibres with vegetable origin, which were later converted into fabrics.

Though there are a number of fibres available in nature, only a limited number can be used to make textiles. Even then, technologists are continuously trying to add a few more fibres such as banana and pineapple at a regular pace into the textile world. However, the essential properties of all such new fibres (tensile strength fineness, pliability, softness) are not entirely favourable in all respects. Therefore, their commercial usage on a normal basis is not fully seen.

Broadly, fibres can be put into four categories depending upon their origin as Vegetable, Animal, Mineral and Artificial.

1.2 NATURAL FIBRES

Vegetable Originated Fibres: They are available in nature in three styles.

1. **Outgrowth over Seed:** The first is fibres from seeds (in the form of seed hair). The cotton (often referred to as 'King Cotton') comes under this style and enjoys the best popularity worldwide.
2. **Bast Fibres:** These fibres are recovered from the inner bark of the stem of the plant. The fibres like jute, flax (linen), hemp and ramie come under this style.
3. **Foliaceous Fibres:** These are recovered from the leaves of plants. Sisal comes under this style.

1.3   SOME OTHER VEGETABLE ORIGINATED FIBRES

1. **Fibres from Fruit:** Fibres can be recovered from fruits; e.g. coconut has got hair-like growth on its surface. When these are removed, they can be used for making ropes.
2. **Fibres from Stalk:** They are actually stalks of plants, e.g. straws of wheat, rice, barley and crops including other crops like bamboo and grass. Even tree-wood falls under this category.

1.4   FIBRES FROM ANIMALS

They are in the form of hair (wool) or filament (silk). Human hair also falls under this category. However, its thickness and surface characteristics prohibit coherence to neighbouring fibres. Hence, it cannot be bundled into rope-like structure even though it is very strong.

1.5   FIBRES FROM MINERALS

In nature, fibres are available in the form of crust. One of the fibres known in this form is asbestos. The basic elements in asbestos are silica, magnesium, lime and oxides of iron. The fibre is very tough, flexible and comparatively longer. Owing to its non-inflammable nature, it has very high potential to be used in fire-resistant and fire-retardant materials or fabrics.

Filaments from Metals: Precious metals like silver and gold have been used in the form of very fine filament in textiles since medieval times. Such fabrics were customarily used by kings and queens and by many rich people. They add excellent ornamenting appearance, thus improving the value of the fabric. Glass, though brittle in nature, also gives useful fibres that have their special application in industry. In many instances, it is used as thermal insulation material.

1.6   ARTIFICIAL FIBRES

These can be further divided into (1) Regenerated fibres and (2) Synthetic fibres

1. **Regenerated:** Cellulose is available in nature in various forms. This cellulose is extracted, dissolved in suitable solvents and then regenerated in the form of fine filaments. Viscose is one such fibre that possesses lustre similar to silk.
2. **Synthetics:** There are large numbers of fibres made artificially by combining chemicals to produce long-chain polymers, which are then extruded in the form of finer filaments. Nylon, polyester, acrylic and polypropylene are some of the popular varieties. These fibres are basically manufactured with an aim to develop special properties. This is done by controlling their ingredients, technical parameters and process parameters during their manufacture.

1.7 FIBRE AS A MOLECULAR CHAIN

Basically, any fibre is a giant molecule which makes it useful for textile purpose. The elements in this long-chain molecule are – Carbon, Hydrogen, Oxygen and Nitrogen. The cotton molecule has as many as 40,000–80,000 atoms. The **Degree of Polymerization** (DP), which is the measure of molecular weight, is the number of repeats of cellulose units that goes into making cotton fibre. The degree of polymerization is the number of monomers (base units) that are polymerized to form the polymer. For cotton, the degree of polymerization varies from 4,000 to 10,000.

During the process of manufacture of regenerated cellulose, special efforts are made to reduce DP to about 350 to make it soluble and ease the extrusion. It also controls the viscosity of the solution, thus minimizing mechanical difficulties during processing. It was found that the length of the cellulose molecule (DP), especially with cotton, is an important factor in deciding the physical properties of a fibre. In general, increasing degree of polymerization correlates with higher melting temperature and higher mechanical strength.

1.8 VEGETABLE FIBRES

Fibres like cotton, jute, flax ramie, sisal and hemp are formed from cellulose. Such fibres serve in the manufacture of cloth and are also used in the paper industry. As mentioned earlier, they can be categorized as **Seed Fibres** (cotton and kapok), **Leaf Fibres** (fique, sisal and agave), **Bast or Skin Fibres** (flax, jute, kenaf, industrial hemp, ramie, rattan and vine fibres), **Fruit Fibres** (coconut-coir) and **Stalk Fibres** (stalks of wheat, rice, barley, bamboo, tree-wood and grass).

Classification of Fibres:[4,6] The most used vegetable fibers are cotton, flax and hemp, although sisal, jute, kenaf, bamboo and coconut are also widely used. Hemp

TABLE 1.1
Degree of Polymerization[6]

Type of Fibres	Degree of Polymerization
Nylon 6	110–120
Nylon 6.6	190–200
Polyester (PET)	100–100
Polyacrilonitrile	2000–2050
Viscose rayon	150–350
Polynosics	700–1100
Cotton	4,000–10,000
Wool	60,000–100,000
Polyethylene (HDPE)	700–1,800
Ultra-high molecular weight polyethylene	100,000–250,000

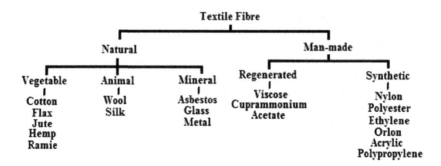

FIGURE 1.1   Classification of Textile Fibres (A)[4,6].

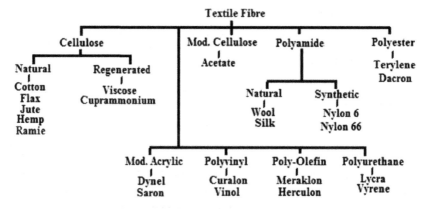

FIGURE 1.2   Classification of Textile Fibres (B)[4,6].

fibers are mainly used for ropes and aerofoil, owing to their high suppleness. They have heat-resistant properties and are also used in the sanitary industry.

1.9   IMPORTANCE OF FIBRE QUALITY

If a fibre has to be useful for its textile purpose, it should have certain physical and chemical properties. Thus, all the substances which look like fibre in nature cannot possibly be considered as textile material. In addition, a fibre must meet the demands of textile users. Also, if it has to be used in actual field applications, it must meet industrial demands.

In addition, fibres and their subsequent forms have to undergo various processing treatments. The conversion of fibre into yarn, for example, involves opening, cleaning, individualizing, and drafting the fibre strand to finally spin the yarn to the desired fineness. Similarly, the conversion of a yarn into a fabric also involves several processes. In all these, the fibres or yarns are subjected to varying stresses and strains. Further, during finishing, several chemical compounds are applied for imparting useful properties such as crease resistance, water proofing and fire retarding. Even after a finished fabric is suitably converted and used by the consumer, the fabric has to withstand washing, laundering, exposure to sunlight, perspiration and abrasion during its lifetime.

The fibres, therefore, must provide for all these, in the form of their strength, sustaining power to processing strain and destructive action of chemicals, sunlight and micro-organisms. For considering the fibres to be useful for their conversion into textile products, there are certain properties which are essential and certain properties which are desirable.

1.10 ESSENTIAL PRE-REQUISITES

Fibre length, strength, cohesiveness and flexibility are the four essential fibre properties.

1. **Fibre Length:** This is the most important property, along with its strength. If a fibre is to be spun into yarn, there needs to be a minimum fibre length of 5 mm. It is because with the traditional spinning methods, it is not possible to spin the yarn below this length. The length alone does not suffice the purpose. In addition, its relation to fibre thickness is equally important. The length to diameter (thickness) ratio needs to be at least a hundred times. Many a time, another indicator called "uniformity ratio" is also used. It is the ratio of the mean length of the fibre (50% span length) to its upper half mean length (2.5% span length), expressed as percentage.

2. **Fibre Strength:** It gives the fibre the ability to withstand the stress-strain caused during its conversion to yarn and fabrics. Certain fibres such as Kapok not only have very short length but also lack adequate strength. The strength is measured as force per unit cross-section when fiber breaks and is expressed as g/d, g/tex or cN/tex. When the two different fibres are to be blended, it is very important to select the components so that they have matching stress-strain curves. When the yarns are spun with this criterion, the two component fibres share the load almost equally.

3. **Fibre Cohesiveness:** The fibres as a bundle finally make the yarn. The ability of the fibres to form the yarn depends upon their cohesive action to hold on to each other. In fact, it is the surface characteristics which give the frictional property to the fibres with which they are able to hold on to the other fibres. The cohesiveness appears in a different form. In cotton, it is due to natural convolutions during their growth. The crimp in the woollen fibres gives them this ability. In man-made fibres, however, this needs to be specially imparted. Viscose is manufactured with serrated cross-section, while polyester fibre is crimped. This is because uncrimped synthetic fibres, as it is, are quite smooth and rod-like structures. In this form, it is very difficult to make them hold together even when twisted.

4. **Fibre Flexibility:** One of the most important operations in yarn forming is twisting. It binds the fibres together to form a yarn having worthwhile strength. During twisting operations, the fibres are strained. The ability of the fibre to try to spring back is flexibility. It makes the fibre pliable.

1.11 DESIRABLE PRE-REQUISITES

Apart from these essential properties, a fibre becomes more useful if it also has some desirable properties. Therefore, even when a yarn can be made with a fibre only having essential properties, a still better yarn can be made when it also possesses some of the following desirable properties.

1. **Fineness:** It is a measure of both the diametric size and linear density of the fibres. In the case of artificially manufactured fibres, the diameter of the fibre is proportional to linear density, except in the case of "hollow" fibres. With natural fibres, the fibre maturity governs the linear density, hence it is customary to define the fibre fineness in terms of weight per unit length with such fibres. The following are some of the units of measure:

   Micron: 10^{-4} cm, especially used in the case of wool fibres to specify diameter. Micronaire: It is another measure of judging both the fineness and maturity of the fibres.

   Micro-grammes per inch: 10^{-6} grammes per inch for expressing linear density. Tex or Denier - grammes per 1000 m or 9000 m, respectively. This unit is specially used for fibres and filaments

   There is a lot of variation in diameter (20–30% for wool or silk) in the case of natural fibres, whereas man-made fibres can be manufactured more precisely (3–5% variation only).

2. **Colour:** The colour of the cotton is judged by two parameters – Degree of Reflectance (Rd) and Yellowness (+b). While degree of reflectance shows the brightness, yellowness depicts the degree of cotton pigmentation. The colour gets affected by atmospheric conditions, impact of insects, fungi, type of soil and storage conditions. There are five recognized groups – White, Gray, Spotted, Tinged and Yellow-stained.

3. **Resiliency:** Fibres exhibit a beautiful property called resiliency. This is disclosed when a fibre is stressed. Here, they try to yield; and when the stress is removed, they try to spring back, i.e. recover their shape and size. Fibres like wool show excellent resilience. During the twisting operation, this property of resiliency becomes very useful when fibres try to regain their original state. Under this condition, a constant pressure is experienced by the inner mass of the fibres, and it helps in holding the inner mass and the peripheral fibres together.

4. **Uniformity:** The artificial fibres can be very precisely manufactured as for their weight per unit length and length itself. But, it is not so with natural fibres. Therefore, in spite of the same growing conditions, natural fibres greatly vary in their size and length. Silk, when formed during cocoon formation, greatly varies in its size. Fibres like cotton have large variation, both in terms of linear density and length. Flax and jute fibres are much stronger natural fibres, but the fibre dimensions greatly vary, thus producing uneven yarn and fabric.

5. **Porosity:** The property of absorbing moisture or any liquid within the fibre arises owing to its porous nature. With the natural fibres, the amorphous and crystalline regions vary in their dimensions. It is the amorphous regions that give the fibre property of absorption. With man-made fibres, the drawing

operation improves the crystalline regions, thus imparting strength. However, their absorption capacity is much limited.

6. **Lustre:** Among the natural fibre, silk is probably the one having maximum lustre. Egyptian cotton has natural silky lustre. Lustre as such is not essential, but when a fibre possesses pleasing lustre, it adds to the appearance. The process of mercerization helps in imparting round shape to the cotton yarns and in improving its lustre. Almost all man-made fibres show far better lustre than many of the natural fibres. However, the glittering reflection of light is not very pleasant. In fact, for this, in some cases, the fibre-dulling process is followed during the manufacture of some man-made fibres.

7. **Durability:** As mentioned earlier, during the conversion of fibre to finished fabric, the fibre, the yarn and subsequently the fabric have to undergo processing, which puts a lot of strain on them. The fibres and their subsequent forms, therefore, should possess durability to sustain these stresses. Even the normal washing/laundering processes or exposure to light and heat require some resistance. The adverse effects of chemicals like alkalis/acids/bleaching agents and bacterial attacks have to be fought against. Fibre durability counts in all such cases.

8. **Commercial Availability:** The supply of regular raw material in the form of fibres is another important criterion. When a particular variety of raw material is not available in plenty, the fibre-processors are forced to switch over to another variety to continue manufacture of the same variety of the goods. The commercial availability is thus another desirable characteristic.

9. **Trash Content:** It is the non-lint material in the bale cotton and is required to be removed before spinning the yarn. Similarly, in all other natural fibres, there is useless and unwanted content. The higher trash content in cotton or in other natural fibres is, therefore, always a problem. Especially in better grade cottons, the high trash content is a huge problem. The trash content in cotton is highly related to leafy vegetable matter, dirt and dust. Normally, the trash content in the different types of cotton varies greatly from 1% to 6–8%.

1.12 SOME TYPICAL BAST FIBRES

These are collected from the inner bark or skin plants. Their main function in the growth of the plant is to support and strengthen the stem. Most of the bast fibres (flax, jute or hemp) are cultivated as herbs.

1. **Flax (Linen)**[1]: It is found in the eastern region of India. In India, it is also known as 'alashi/javas'. Long time ago, it was reported that flax was grown in Egypt. It is an annual bushy plant with a height of around 1.4–1.6 m. The plant has small green leaves and stands erect when fully grown. The flowers have colours ranging from pale blue to bright red. The round fruits inside contain shiny brown seeds. The fibre bundles inside the stem appear like 'blonde hair'.

   Physically, the fibre is soft, lustrous and stronger than cotton. However, it is comparatively stiffer (less flexible). The better varieties in flax are used to make linen fabrics (damask, sheeting, etc.). The coarser fibre varieties are used for making twines and ropes.

Whereas the fibres are useful in making banknotes, the seeds have high nutrition values and contain 'Omega-3'. The seeds are very useful for heart patients as they control cholesterol level. Recently, it is found that the seeds also have anti-cancer characteristics.

2. **Jute**[1]: It is a long, soft and shiny fibre. It is possible to spin very coarse and strong threads from this fibre. It is owing to this property and its corresponding application in the industry that jute is the next largest grown natural fibre, almost next to cotton. It is mainly composed of cellulose and lignin. It has high tensile strength, low extensibility and better breathability. Like cotton, it can be blended with both natural and synthetic fibres and can be easily dyed. It is often treated with caustic soda to usefully improve its softness, crimp, pliability and appearance.

On the other hand, jute has poor drapability, crease resistance. The fibre is also comparatively brittle. However, enzyme treatment can reduce its brittleness and stiffness. When exposed to sunlight, it turns yellowish. The treatment with castor oil can reduce this tendency. When the fibre is dyed, it becomes brighter. However, unlike cotton, its strength decreases when it becomes wet. In this condition, micro-organisms can easily attack the fibre. To improve its dye affinity, it is often treated with enzymes.

The fabric made from jute is popularly known as 'hesian cloth' or 'burlap. As the fabric is very rough and tough, it is used in making sacks or 'gunny bags', commonly used to store food gains or wrap or cover the material.

Normally, jute is grown during the rainy seasons. The stem is cut. It is then retted and the fibres are recovered. During recovery, the non-fibrous matter is then scraped-off and thrown away. It is very widely cultivated in the delta of river 'Ganges'. There are two popular varieties – 'Tossa' jute and 'white' jute. The former is softer, silkier and stronger than the latter. The jute is largely grown in West Bengal (India) and Bangladesh. The common bale size for jute is 180 kg.

With the advent of polyethylene, the gunny bags are presently made from this synthetic fibre (in the sheet form). In spite of this, jute is very popular in the manufacture of many industrial products such as paper and typical textile hand-made articles. With the development of non-woven textiles and composites, jute has been found in many geo-textile applications as the jute cloth has very good protection against soil erosion. Jute is bio-degradable and this is advantageous over any other synthetic material. Therefore, it is popularly used as containers for planting young trees. The jute fabric also acts as an excellent seed protector and weed controller.

The research is being carried to fruitfully blend jute fibres with other synthetic fibres, and this has successfully led to making ropes and twines with special properties. It has also been made possible to use fine jute threads as 'imitation' to natural silk. The use of the jute fibres in the paper industry has saved many trees from being cut – useful ecological aspect. It is extensively used in the manufacture of decorative woven and tufted carpets and mats.

With the present fear of partial destruction of ozone layer, more attention is given to protection from harmful UV rays. Apart from sound and heat insulation, jute is capable of providing protection against these harmful rays. In addition to these very beneficial properties, jute also possesses anti-static properties. Therefore, it has become very popular in home decorations.

Jute leaves have high nutritious values (iron, calcium and vitamin C) and is an excellent anti-oxidant. Dunlop mats with jute cloth covering is a very popular bedding material. Also, owing to its thermal insulation properties, there has been increasing use of jute cloth in automobiles.

3. **Hemp**[1]: It is another bast fibre, which is soft, durable. The fibres are obtained from the plant and surround the interior woody portion of the stalk and inside the outer bark. These are mainly used for industrial and commercial purpose. The fibres recovered from the plant usually vary in length. Basically, these fibres give strength to the plant.

   The colour of the fibre also varies from off-white to brown or blackish green. As compared to cotton or even flax, the fibre recovery is comparatively better. Further, the pesticides and herbicides required to be used during the growth are much less. Earlier, hemp was used to make canvas and many types of cordages, basically owing to its strength.

   Nowadays, hemp is being widely used for industrial purposes such as paper-making, bio-degradable plastics and health food. Poor quality hemp is used for medicinal purpose. Another purpose for which hemp is grown is in making marijuana (a drug). Some research work has been carried out to use hemp in making fuel, and it has met with some success.

   As for its textile use, hemp is often blended with cotton or jute or silk to make apparels and furnishing fabrics. In construction, it is popularly mixed with concrete as it strengthens the structure. Automobile panels are made of a combination of fibre glass and hemp fibre.

   Seeds have high nutritious values. As they contain many useful body minerals, amino acids and fatty acids, they are used in making many food items such as cakes, protein powders and even milk (like soya milk). The oil with its rich ingredients is known to improve memory and strengthen brain cells. The oil also forms an important ingredient in some typical oil paints.

4. **Ramie**[3]: It is a typical bast fibre, basically grown in the tropical region and plant normally reaches a height of 1–1½ m. The leaves are broad and heart shaped. The white leaves have silvery hairs on their underside. The white ramie is popularly known as China grass. The green ramie is also grown.

   Under ideal conditions, it grows very rapidly. The degumming process involves chemical treatment. After the flowering, the fibre content in the plant is the best. At this time, the harvesting starts. The stems are cut very close to the roots. The core is broken and the cortex is stripped. When the bark is soft, the decortication starts. The degummed fibres are recovered. Efforts are made to restrict the loss during degumming. Most of the bark, gum and pectin is removed during scraping of the cortex. When the residue is washed, it leads to all spinnable fibres.

Ramie is a very strong fibre in wet condition. It has the ability to hold shapes. It does not wrinkle and possesses silky lustre. It is often used in blends. It is similar to flax in absorbency, density and microscopic structure. It is however, less durable and does not dye well. The fibre is stiff and brittle. It shows poor resiliency and elasticity. Therefore, in spite of its high strength, it has limited use in textile. The yarn is quite hairy because of the lack of coherence, and the spinning and weaving of the fibre is difficult due to its brittleness and lack of elasticity.

As the cloth has toxic properties, in olden days it was used as 'mummy' cloth in Egypt to protect the dead bodies from bacteria and fungi. It was also used for making shirts and gowns.

Ramie is used for making industrial sewing threads, packing material, fishing nets, filter cloths, upholstery and furnishings. Shorter fibres are used in pulps for paper industry. Soon it will enter into the automobile industry as bio-plastic material.

5. **Kenaf**[3]: It is a typical bast fibre. The fibre resembles jute. In India, it is popularly known as 'ambadi or Deccan hemp'. It is also grown in China and Malaysia.

    The fibres present in outer stem are coarser, whereas those from inner stem are comparatively finer. Within a year, the plant grows to a height of almost 2–3 m. The leaves are typically long and have lobes at the base. During flowering, the colours of the flowers vary from white-yellow to purple. When the flowers wither, they leave behind a pod or capsule. This is actually the fruits and it contains many seeds. The leaves are normally used as a vegetable. The leaves can also be used as food for animals.

    The fibres obtained are coarser and, hence, are mostly used to make ropes, coarse cloth, in paper industry and for making boards. It is used with resins in plastic composites. The juice is beneficially used as coolant in drilling fluid. The fibre has a special use, in that it offers an excellent material for molding it into grass mats for containers and for making instant lawns.

    Being basically a cellulose material, it is very popularly used in the paper industry. The sheets of paper thus made are quite strong for printing. As it contains comparatively less lignin, less energy is required to turn the kenaf fibre into pulp. The fibre has natural dull colour and hydrogen peroxide is used as bleaching agent to remove it. It is very economical to grow the plant as it requires less pesticides and water. When grown on a large scale, about 8–10 tons of fibres can be easily obtained per hectare (approx. 20-24 acres).

6. **Banana Fibre**[3]: Though the plant is mainly grown for their fruits, the research carried out led to a very useful fibre that can be recovered from the 'pseudo-stem' of the plant. The stem usually grows up to a height of 3m. The fibres are normally rough and coarse in nature very similar to jute fibres. The quality of the fibres recovered from the stem is usually better. Like Kenaf, the fibres from outer stem are comparatively coarser, whereas, those from deep inner fibres are much softer. A softening treatment brings about some favourable change to make the fibre spinnable. Depending upon the length of the stem, it is possible to have a fibre length up to 2½ m.

The stem shoots are initially boiled. Then follows the mechanical extraction. The fibres are partially bleached and dyed. The coarser fibres are used for the manufacture of rugs and table cloth, whereas from finer ones, kimono and kamishimo are made in Japan.

Though very light in weight, the fibre is strong having strength varying between 29 and 30 g/den and its breaking elongation is around 6.5%. The moisture-regain value is high (around 13%). The speciality of this fibre is that it both absorbs and releases moisture quickly. The fibre is bio-degradable and eco-friendly.

Banana fibre is used in apparel garments and home furnishings. Being little rough in nature, ropes, mats and some composites are also made from it. It is being used in building construction, geo-textiles and even in sound engineering for sound-proof boards. Hand-made papers are made from the pulp of banana fibres. The paper made from bark is often used for artistic purpose.

1.13   FIBRES FROM LEAVES

1. **Sisal**[4]: The fibre is a little stiffer and hence mostly used for making heavy ropes, thick twines and dartboards. More than 100 years ago, Sisal was found to be grown in a few American and African countries. The recovery of the fibre and then the actual spinning started much later. Presently, Brazil is a large producer of this fibre.

   The plant is mostly grown in the tropical region. The leaves are long and sword-shaped. They have sharp biting teeth. The plant bears numerous leaves during its life span. The percentage of fibres recovered from the leaves is comparatively very small.

   The leaves are washed with water. Their surface is then lightly scraped-off (decorticated) and the fibres are recovered. They are dried, lightly brushed and packed into bales. The drying of fibre has to be done carefully as the moisture level in the fibres controls their properties, This, in a way, improves fibre quality. In another method, water is used to separate the fibres. The fibres are combed and sorted out in various grades.

   Sisal can be used in making rough and strong cloths. It is useful in the manufacture of wall coverings and carpets. Owing to its strength, it is proved to be a good binder. It is the basic material for making paper, filters, geo-textiles, mattresses and handicrafts. It is also used in composites in place of fibre glass and asbestos. The fibre has higher abrasion resistance and hence, it is advantageously used in cordages and carpets. Sisal carpets easily carry the static away. As the dust and impurities are not trapped, only vacuum cleaning is required to clean carpets. Depending upon the surrounding relative humidity, sisal can either absorb or release the moisture. It is a very good food for honey bees. However, the honey is bit darker and does not have a pleasant flavour.

2. **Fique**[3]: It is a national fibre of Columbia. It is available in nature as the fibre in the leaves of the fique plant. The leaves are dark green and do not have spikes. The leaves are weak and hence they droop downwards. Many ethnic products are made from this fibre, owing to its heat-resistant properties, it is

beneficially used as protector while holding the hot tea or coffee cups.

After the fibres are extracted, they are normally used in making ropes and hammock strings (swing chairs). It is also popularly used as a packing material in agriculture. The sacks with tight structures are used for food grains, whereas those with a more open structure are meant for fruits or vegetables. The fibre being quite strong in nature may also be used in the manufacture of sailcloth, mattresses, shoes and handicrafts.

The fibre in the pulp form is used by the paper-making industry. In the liquid form, it is used in making fungicides and soaps. Special types of beverages can be made using fique. The woody portion (stem) is mostly useful in building construction. The fique bulbs are very good in treating boils. The bulbs also have edible value and hence after boiling are used in making food.

3. **Pineapple Fibre**[3]: The fibre is again a leaf fibre. After cutting the leaves, their surface is scrapped, The fibres from inside are gently pulled out. The strands of fibres are again hand scraped and are converted into a long continuous strand. Then follows an industrial process of converting the strand into a suitable and useful textile product. The general look of the fabric made from fibres resembles canvas and can be dyed, printed and also treated to develop different textures.

The strength and fineness are nicely combined to make a typical fabric – 'Barong Tagalog'. The fibre is also used for making table spreads and mats. The fabric is usually lightweight but stiff and normally gives leather-like appearance, and hence is used in making small purses, bags or even shoes.

1.14 FIBRES FROM FRUITS

1. **Coir**[4]: India (coastal part of Kerala) produces a large quantity of coir fibres. These fibres grow on the hard internal shell of a coconut. With an increase in fibre maturity, there is an increase in lignin deposits, which makes the cells hard. In the early stage, when the fruit has not ripened, the fibres are whitish, smoother, finer and weaker. As the fruit ripens, the coir possesses less of cellulose and more of lignin. This increases the abrasion resistance of the fibre. The fibres also become stronger and stiffer (less flexible).

The products made from the yarns are mostly for rough and sturdy use. Thus, the fibre is used for making mats and ropes. Being mostly grown around the sea-shore, the fibres are comparatively more waterproof. They are, therefore, used in fishing nets. Green coconut gives pliable white fibres. The brown fibres on matured fruit are soaked in water to soften them, and the longer and shorter fibres are separated (wet-milling). The fibre mattress thus obtained contains dirt and other waste matters, which are removed. The fibre is then dried and packed into bales. With some moisture remaining within the fibres, the fibres show elasticity. This stage is very suitable for twisting them into ropes. In the villages, hand twisting is convenient and hence more popular.

The fibres, if required, can be further bleached and dyed. Brown coir is used for doormats, brushes, and floor tiles, in sack material and rope making. The fibre mattress when needle punched can be used for padding in bed-mattresses and are excellent material for controlling erosion. With rubber sprayed on these fibre mattresses, it forms excellent material in latex form and is used in upholstery padding in automobiles. The fibre sheets are good insulator.

1.15 FIBRES FROM STALK

1. **Bamboo**[3]: These plants truly belong to the grass family. The plant stem is hollow and the vascular fibre bundles are scattered throughout the inside stem. The soil condition and climate around govern the plant growth rate. Some of the large bamboos form timber. The bamboos are harvested when their strength is high and the sugar content is low, as this protects it from pest infestation. The woody portion containing vascular tissues (sap wood) of the bamboos is reduced by the 'leaching' process. One of the processes to convert bamboo into fibrous matter by mechanical means is similar to the one used with flax. In this, stalks are crushed and given enzyme treatment. In other process, the bamboo fibres are broken down (similar to rayon manufacturing) with chemicals and extruded through spinnerets.
The fabric from bamboo is softer and shows high absorbency and anti-microbial property. It makes almost 99% of Panda's diet. The fibres are used in the paper-making industry, especially high-grade hand-made quality. Being hollow, bamboo is very popularly used in musical instruments. Bamboo filters are used in removing salt from salt water. It is used in construction as reinforcement to concrete. However, in this case, it should be treated to resist insects and rotting. In Japan and China, laminated floorings are made from bamboo fibres.

1.16 ANIMAL FIBRES

1. **Wool**[1]: Most of the animal fibres are protein based, e.g. silk wool, cat-gut, angora, mohair and alpaca. The wool is obtained from sheep hair. The famous cashmere wool is obtained from cashmere goat. Though 'wool' really means hairs obtained from goat, the hairs from other animal – are given appropriate names like 'mohair' wool, camel wool, 'angora' (rabbit) wool and 'alpaca' wool. The undesirable hair part of the fleece which is very weak, stiff and short in length on the ship is called 'Kemp' (It does not pick-up dye and reduces the quality of the wool). The proportion of kemp to wool varies from ship to ship, depending upon the breed.
Wool hair has some special characteristics such as crimp, elasticity, special texture and handle. It is a staple fibre and has typical scales on its surface. This phenomenon, together with crimp, gives clinging power to wool and makes it hold on to the other neighbouring fibres. The crimp varies with fibre fineness. The finest 'Merino' wool may have even 100 crimps per inch (around 40 crimps/cm) whereas very coarse may hardly have 2–4 crimps per inch. Wool is

hygroscopic in nature and can absorb up to one-third of its weight without feeling wet. The fibres also absorb sound. It is generally creamy white but in some instances, brownish fibres are also seen.

When wool fibre is sheared from the sheep, it contains a sizable proportion of grease, dirt, vegetable matter and other impurities. After shearing, the good fleece is separated from unwanted broken, bellies and locks. The grading is also based on the fineness of the fibre. The fineness is generally defined in 'microns (10^{-4} inch). There are other considerations like length, tensile strength, colour and comfort. Even then, the fibre diameter is the most important criterion which is accepted to differentiate the quality. The typical merino wool (3–5 inch) is very fine (15–20 microns). In general, wool finer than 25 microns is used for making garments. As against this, very coarse wool fibre (>32 microns) goes into making outer wears and rugs. Though finer wool is softer, the coarser grades are more durable and less prone to pilling. The best merino wool coming from Australia and New Zealand is very soft and finer (around 16–17 microns).

a. **Wool Processing - Shearing:** As compared to wool produced in Australia, China and New Zealand, India has a very small share of wool production. The processing of wool starts from the first operation – shearing of fleece from sheep's body. Wool shearing is a very skillful operation, and has to be done very carefully. The wool obtained from shoulder and sides is fine, strong and uniform, whereas that obtained from neck, back, belly and forelegs is short. On an average, fibre shearing is done at least once a year. Normally, in countries where there is severe cold in winter, the shearing operation is avoided.

A shearing blade very similar to normal scissor is used. With this, it is possible to cut the wool close to the goat's skin. Nowadays, machine shearing is very common. The power-driven cutters give a smooth cutting without much wastage.

b. **Grading:** Even when the sheep belong to the same breed, the flock of the fleece obtained often varies. After shearing, therefore, it is necessary to se-parate the sheared stock into classes according to their characteristics. This brings an uniformity within a class. There are expert 'Classers' who classify the given fleece as a whole.

Equally, interesting is to note typical non-wool substances associated with wool-fleece – '**Yolk**' which is an animal-originated extraneous material. It contains greasy matter and suint (sweat). Whereas suint is water soluble, grease has to be removed in scouring operation. Either embedded in the grease or otherwise, there is a certain quantity of '**sand**' and '**dirt**'. There are other vegetable entanglements – '**burrs**', mainly arising owing to nature of pasturage (grazing land). All these substantially affect the yield of clean wool. Finally, the wool is graded by the experts, who take into consideration the fineness, length, colour and other important properties. Thereafter, the graded wool is packed into large sacks and sewn to form bales of approximately 300 lbs (135 kg).

As the wool is being prepared for sale, its price is assessed by the buyers de-pending upon its fineness and length. The number given to wool (e.g. 80^s wool) is the count to which it can be beneficially spun. Here, the number 80 is the worsted count (no. of 560 yards in 1 lb). In woollen trade, the count is 'no of 256 yards in 1 lb'. Typical merino wool usually lies between 80^s and 100^s, whereas the wool used for carpets is very coarse and can be spun to count less than 40^s.

c. **Scouring:** As mentioned earlier, the raw wool (after shearing) contains lots of contaminations. In most of the cases, these impurities are present up to 50% by weight of raw wool. Further, there is some proportion of greasy matter and suint (sweat). In general, finer wool like merino wool contains a comparatively higher proportion of these impurities than coarser wool.

The wool is gently washed in a warm water tank, containing alkalis and some detergents. There are four or more such tanks and through each, the wool is propelled gently and squeezed. Finally, the wool is rinsed with clean water and dried, allowing 20% of water to remain. In other recent methods, the grease dissolving solvents such as paraffin or petrol are used to wash the wool. The machinery used for processing woollen (coarser varieties) and worsted (finer varieties) differ in their processing sequence. The worsted spinning is very close to cotton spinning. The 'worsted' is long-staple fibre, strong and combed wool yarn; whereas the 'wool' is soft, short-staple carded woollen yarn often used for knitting.

With a small amount of twist, the fibres in the woollen yarns are loosely held. Therefore, the yarns are fuller and thicker. As such, short-staple wool is used for making these yarns. As against this, the worsted yarns are made from fine long-staple wool. They are finer and with a full twist are firmer, smoother and stronger.

Though the scouring removes greasy matter and suint, it does not remove vegetable-originated matter like twigs or burrs. These are destroyed by treating wool with dilute solution of strong acid like sulphuric or hydrochloric acid. The process is known as 'carbonizing' process. Sometimes, heavy crushers are used to powderise these matters, which can then be removed in the subsequent carding process.

A brief outline is given below for the process of worsted and woollen spinning in Figs. 1.3 & 1.4.

Woollen yarns are often spun from a mixture of wool – new and recovered. Some other fibres may also be incorporated.

Worsted Processing

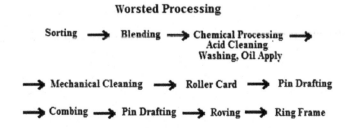

FIGURE 1.3   Machinery Sequence for Worsted Processing[1].

Woollen Processing

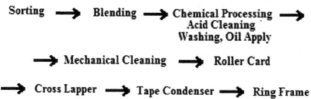

FIGURE 1.4   Machinery Sequence for Woollen Processing[1].

    d. **Wool Carding:** The woollen card is a special card. Like cotton card, the object of this process is to disentangle the matted mass of wool fibres. The carding process is almost repeated in a sequence in three parts – scribbler, intermediate and final carder. Thoroughly carded wool in the form of a thick sheet, almost resembling a blanket, emerges from the final machine.

    This blanket sheet is split into a number of ribbons, quite thick in size. They are passed through pairs of rubbing aprons in a machine called 'Tape Condenser' The ribbons thus receive a false twist sufficient to hold the mass within itself. The loosely twisted mass called "Slubbings" are rolled into a loose package. Traditionally, these packages are directly taken on spinning machine to spin a soft, fluffy woollen yarn.

    e. **Worsted Yarns:** In this, the wool is immediately dried after scouring and sent for carding process. As in cotton card, here the thin fibre sheet delivered is condensed into sliver. It can be rolled up into balls or simply coiled into cans. Long length wool, however, is gilled. In this machine, as the fibre strand is passed through pairs of rollers, it is subjected to a combing action by the rapidly moving rows of pins. The speed of the pin rows is faster than the speed of the strand through the rollers. The fibres are thus straightened and aligned to the material axis. The idea of this kind of alignment is a typical feature of worsted manufacturing and basically resembles a sliver drawing (drafting) in draw frame.

    The purpose of wool combing is again similar to cotton combing. In this, apart from removing the short fibres, the main strand of long fibres is further aligned. Again like cotton combing, in worsted processing, it is common to employ gilling even after combing. The wool in this untwisted form is called 'top'. It is further drawn-out to form a roving very similar to the one in cotton spinning.

    There are different ways of manufacturing worsted yarn. In old days, it was 'Cap Spinning' which was used to produce cops. The rubbing action around the outer surface of cap, however, made the yarn more hairy. Flyer spinning is another method and is found to be useful in making a smooth yarn. Ring spinning is the high production method to produce the yarn. Little finer varieties of worsted yarns are preferably spun on ring frame. Very old method of mule spinning used to give softer, fuller and more even yarns. However, the process was intermittent and hence slow. Further, the machine required a huge space and the carriage carrying spinning spindles were used to move to and fro. Hence, the tenter also had to move along with the carriage. With the higher production trends, this machine, in no way, could compete the ring spinning. On an average, in 1 lb of wool with a very fine count of 125^s (Woollen hank = No. of 560 yards in 1 lb), the length of the yarn can be as high as 70,000 yards. Both the number of processing stages and the higher price of the raw material, make the worsted fabrics quite costly. As a general rule, the finer wool has more no. of scales and crimp in a unit length of the fibre. The fabrics made out of such wool are quite firm but lighter in weight.

As mentioned earlier, the worsted yarns are much tighter than woollen yarns. Therefore, the amount of twist put into worsted yarns, especially those made for suitings or gabardines gives them harder texture. Owing to stronger yarns the fabrics are more weather-resistant. However, there is another category of worsted yarns where the yarns are soft twisted. The fabrics, thus made are soft, warm and lightweight. Low-twist yarns are also made when later they are milled or felted.

f. **Bleaching & Dyeing:** In bleaching, the wool is discoloured to make it appear whitish. As a general rule, when the fibre is to be dyed later into dark shades, it is not necessary to fully bleach the wool. In this case, partial bleaching is carried out. However, when the yarns and therefore the fabrics are dyed in brighter and lighter shades, the woollen material is required to be fully bleached. The bleaching is usually carried out in either yarn or fabric stage.

Earlier, wool used to be bleached with sulphur dioxide. But the bleaching effect was never permanent. As chlorine-based bleaching is not suitable for wool, hydrogen peroxide is used to bleach the wool. Maintaining pH around 8.5–9.0 and temperature around 55°–60°C, the bleaching process is carried out.

Temperature and time are the two important factors in dyeing. During dyeing, the dye molecules develop a special chemical bonding with wool-fibre molecules. The artificial dyes are produced to obtain a broader range of colours. The dyes are also made more stable with a view to develop washing fastness. With acid dyes, wool is dyed in the hank form. Direct dyes are also used on wool and they are water soluble.

g. **Wool Felting**: The scales on the surface and the elasticity are mostly responsible for wool shrinkage. When a woollen garment is washed, even with due care, it shrinks. A rain shower loosens the fibres and causes relaxation shrinkage. Felting shrinkage, however, takes place in a different way. Felting occurs when woollen fabric is subjected to mechanical action when it is wet. The fabric becomes thick and the fibres become closely packed. The yarn pattern loses distinctness and the fabric becomes much less elastic. The wool fibre felts more easily with softened fibre tips or hardened fibre roots. Generally, any form of rubbing would easily felt the wool. So also, a loosely knitted structure (e.g. a fluffy sweater) would felt more easily than a tightly woven flannel or worsted trouser.

The natural felt is a fibre mass, held together without being woven. Whereas, some special felts are made by simply weaving the fabric and then felting them. In the blankets, the felting causes the weave to become almost invisible. The wool is basically used in blankets and rugs, carpets, felts, upholstery and for insulation purpose. It has the ability to absorb both odor and sound, hence used in heavy machinery and sound equipment. The cloth diapers are made of two types of linings – inner one being hygroscopic and the outer one hydrophobic. When felted and treated with lanolin, the wool becomes water resistant and yet air permeable. In this state, it shows anti-bacterial tendencies. Basically being protein, wool is used as fertilizer where there is a slow release of nitrogen and ready-made amino acids.

'Shoddy' or recycled wool is made by cutting or tearing apart the old, used fabrics. This reduces the length of the fibre appreciably, and hence the fabrics made out of this wool are inferior to those of virgin wool. Recycled wool may be mixed with virgin wool or any other fibre such as cotton to improve its functioning. These yarns are generally used as weft.

As such washing of woollen clothes is done by a process known as 'dry-cleaning' when petrol is used. The new technology – 'super wash' wool (washable wool) has been introduced. In this, with acid treatment, the scales are removed. In another process, the fibres are coated with polymer to submerge the scales. This avoids shrinkage during normal machine washing.

Today, however, owing to the rise in the popularity of synthetic fibres, the demand for the 100% woollen garments has considerably reduced, though in severely cold countries, it still is in great demand. Nowadays, it is more popularly used in a blended form with polyester.

2. **Mohair**[3]: The fibre is obtained from angora goat. The yarn and the fabric woven from it resemble silk. The fibre diameter of mohair ranges from 25 to 40 microns. It is a 'keratin' (protein) fibre. Mohair has scales, but they are not fully developed as in the case of wool. Hence, it does not felt like wool. For the older goats, the hair is coarser in diameter. The shearing is done twice a year – approx. 5–6 kg yield. The grease and other impurities including vegetable matter are removed in similar manner as with wool.

It is one of the oldest and luxurious fibres, both durable and resilient. It reflects high lustre and sheen. The fibre is warm and has insulating properties. It shows exceptional dyeing properties and owing to its good moisture-wicking property remains cool in summer. It is both flame and crease resistant.

It is used in folding roofs of convertible cars. There are many other applications – in scarves, winter hats, suits, sweaters, coats, socks and home furnishings. It is also found in carpets, wall fabrics, craft yarns and may be used as an economical substitute for furs. Having resemblance to human hair, it is used as wig for high-grade dolls and customized dolls.

3. **Alpaca:** It is a domesticated species and resembles a small llama. The animals live in herds and often graze on the high mountains (above 4000 m height). The most famous fibre comes from Peruvian alpacas. Like other animals, the fibre is sheared from the body of the animal. The fibres vary

greatly in colour and they are commonly known as 'camelid'. In olden days, the animal was slaughtered for its meat, but today it is a protected animal.

Alpaca is a natural fibre possessing silky lustre. In feel, it is warmer but not prickly and shows low allergenic tendencies. With lanolin around the fibre, it repels water. It is much softer and glossy like human hair. The process of preparing yarn is similar to that of wool. In Australia, depending upon, its thickness and trashy matter, the fibre is actually priced.

Similar to wool, it is used in knitting and other woven items. These include blankets, sweaters, hats, gloves, scarves, socks, coats and materials in bedding.

4. **Silk**[1]: Out of almost 30 countries producing silk, China's share is almost half. In India, about 12–15% of the world's share of silk is produced. The cultivation activity is termed as 'sericulture'. The fibre/filaments are recovered from the cocoons formed by the silkworms. Several other types of moths prepare such cocoons, but the thread obtained from that of silkworm is far superior. The silkworm is typically grown on 'mulberry' trees. The cocoons contain a large number of layers of silk thread, and they vary in their softness and structure. The larvae before starting to build up cocoon, seeks some place of hiding or creates camouflage to safeguard the cocoon. At the time of cocoon formation, two primary glands under the head of larvae secrete silk threads (fibroin), which are bound together by sericin which is a protein in the form of a glue.

Recovering Silk from Cocoon: The insects after the completion of their growth inside the cocoon, try to break open the cocoon by secreting some fluid to soften it. It is just before this stage that the cocoons are gathered and boiled in water to soften the gummy matter and also to stop the inside worm from breaking it open. When softened, the expert-reelers slowly unwind the silk thread from the cocoon. It is really a skilled job to reel the continuous single thread from one cocoon. The length of the thread from single cocoon varies greatly from 300 m to even 1000 m. When reeled without any break, it gives a much stronger cloth when subsequently woven.

General Information: The silk fabric is quite luxurious in appearance and is mostly worn by rich families owing to its texture and lustre. The silk in India is referred to as 'pattu' or 'resham'. The silk is traditionally hand-woven and dyed. In highly priced sarees, even silver threads are used for ornamentation. There are other varieties – 'muga', 'eri' and 'pat'. They are produced from silk worms prevalent in different areas.

Properties: The silk fibre has a triangular cross-section with rounded corners and in fact, it is these rounded corners that are responsible to give the silk its typical lustre. With some species, however, the cross-section can be even crescent. The fibre composition is fibroin: 70–80% (actual fibre content); Sericin: 15–18% (the binder or glue), fats, ash and colouring matter about 2–3%. The finest silk has a diameter varying from 5 to 10 μm; whereas with very coarse varieties of Tussah silk, it can go up to even 60–70 μm. The silk has smooth and soft texture and the fibre-filament is not slippery. It has tenacity which varies from 4.2 to 4.6 g/d. However, in wet condition it loses about 15–25% of its strength. When silk is burnt, it gives a smell similar to when human hair is burnt. However, with the flame removed, it stops burning. As against this, artificial silk smells like plastic when burnt. The silk has quite a high moisture regain (11%). However, it has poor elasticity and when elongated, it remains stretched. If it is exposed to sunlight for a long time, the filament becomes weaker.

As it is a poor conductor of electricity, it is susceptible to the accumulation of static. In unwashed condition, due to the relaxation of macro-structure, the fabrics like chiffon show shrinkage even up to 8%, though this shrinkage is not due to any molecular deformation. Hence, before garments are made, it is customary to wash the cloth.

The silk is moderately resistant to most of the mineral acids. However, sulfuric acid dissolves it. The strong alkalis weaken the fibre. The tensile strength acquired by the silk is due to strong hydrogen bonds. These bonds do not break even on stretching. On absorption of perspiration, silk fibre turns yellow. Having much higher absorbency, silk is very comfortable in warm weather conditions. During cold weather also, being bad conductor of heat, it retains body warmth. Silk has a variety of uses. The shirting cloth, blouse material, formal dresses, high fashion clothings, pyjamas, robes, skirts, ties and kimonos are preferably made of silk. Its soft lustre and excellent draping ability make it most suited for furnishing fabrics, upholstery, wall coverings, curtains, beddings and wall hangings.

Silk has been put to many industrial and commercial applications. Much earlier, bullet-proof vests were made of silk. It was popularly used in parachutes. When outer sericin coating is removed, silk becomes softer and then can be used as non-absorbent suture threads. When used as an underclothing, it helps in reducing itching. Its bad electrical conductivity makes it the best insulator in electrical trade.

1.17 MAN-MADE FIBRES

Some of the man-made fibres are made from natural cellulose. These are termed as 'regenerated' fibres; whereas the other category is purely synthetic fibres which are manufactured from synthetic material after polymerization. The first category includes fibres such as viscose, modal, Lyocell and cuprammonium, whereas in the second category, there are fibres such as nylon, polyester and acrylic.

1.17.1 REGENERATED FIBRES

1. **Viscose (Rayon)**[7]: As mentioned earlier, it is regenerated by dissolving cellulose available in nature or the one recovered from other sources. It is then reformed and spun as a filament. Owing to its resemblance to natural silk in its lustre, it is also known as 'art silk'.

   The cellulose is dissolved in caustic soda and is then treated with carbon disulfide to form viscous liquid 'cellulose xanthate'. The main source of cellulose here is wood pulp. The contaminants are removed during washing with water. However, many hazardous effluents are set out, and many of them have detrimental effects on the surrounding environment. The solution after filtration is again dissolved in caustic soda and allowed to ripen for a certain period. It is then filtered to remove the remaining impurities. This viscous solution is then extruded through spinnerets into coagulating bath containing sulfuric acid. As soon as the liquid-jets from the spinneret enter the bath, they solidify to form the filaments. The subsequent drawing operation stretches the newly born filament. This stretching or drawing operation is done for two distinct purposes: first, it reduces the filaments to the required fineness and second, it orients the molecules within the filaments. This improves the tenacity of the filament. Usually a single filament constitutes many finer filaments depending upon the size of the spinnerets. Such fine filaments are then bundled as one main filament. The viscose thus manufactured can be used in its filament form or can be cut stapled to a much smaller length depending upon its ultimate usage.

Normal viscose, especially when wet, is comparatively weaker. This has led to the manufacture of high wet modulus quality (HWM). It has got improved strength even when wet. These rayons are also known as 'polynosics' or 'modal'. High tenacity rayon is another modified version which gives substantial improvement in tenacity. It is specially manufactured for industrial use. Another variety resembling viscose is cupra-ammonium rayon. Here, the only difference is that copper and ammonia are combined with cellulose. However, it is not produced much, as during its production; there are many detrimental discharges that pollute the environment.

Viscose has circular but indented cross-section with striations along its length. As against this, the HWM and cupra-ammonium fibres are round in cross-section. The viscose filament yarn is composed of finer filaments ranging in number from 80 to 500, with their denier varying from 40 to 3000. The staple fibres are made in the range of 1.5–15 denier. The fibres are also crimped to increase their coherence and bulk. On many occasions, the pigments are added (delustring) to reduce its glaring lustre.

As artificial cellulosic fibre, viscose has comfort properties, texture and a feel like silk and cotton. It can be dyed in wide ranges. It is smooth and cool and shows higher absorbency. It shows poor insulating property. But this makes it suitable as apparels wear in hot and humid climatic conditions. However, durability and resistance to creasing are very low, especially in wet conditions. Therefore, clothings can get deformed during washing. So also, it has low elastic recovery. As against this, HWM rayon shows improvements in these characteristics. Thus, when HWM fabric can be machine washed, normal viscose fabrics need dry washing.

2. **Modal**[3]: It is made from cellulose which is recovered from beech-wood tree. Like cotton, it is hygroscopic and takes-up the dye easily. Its washing fastness with dyed fabrics, especially with warm water is really good. The fabrics made from this variety of rayon are smooth, soft and are resistant to shrinkage and fading. Modal fabric is quite resistant to mineral salts and prevents sticking of mineral deposits when washed with mineral water. It, however, requires ironing at moderate temperatures.

Being soft and smooth, modal fabrics are used for towels, bathrobes and bedsheets. It can also be blended with other fibres to modify the usage.

3. **Lyocell**[3]: It is another variety of regenerated cellulosic fibre made by dissolving bleached wood pulp. It is also manufactured under another brand name 'Tencel'. Lyocell is formed by 'solvent spinning process'.

The wood pulp is dissolved in a solution called 'dope'. This dope is forced through spinnerets to form filaments. The filaments are washed when chemicals used are retrieved back. These chemicals can then be purified and recycled. During the processing, there are no by-products and the process is more eco-friendly. However, a large amount of energy is required and the solvent itself being the by-product of petrol, it happens to be costly. In finishing process, lubricants are applied. These usually are in the form of soap or silicone. It helps in smoothening subsequent spinning processes.

Tows (bundle of filaments) are then formed and taken to crimper. The crimped fibre is then carded to separate the fibre strands. The strands are then cut to the required staple and baled. The fibre is either blended or the yarn is spun from 100% Lyocell fibre. The yarns are either woven or knitted and the fabrics are given variety of finishes.

The Lyocell fibre is soft, absorbent and extremely strong when wet or dry. The fabrics can be machine washed and have excellent resistance to wrinkling. They drapes well and can be dyed in many colours. However, it is a more expensive fibre. Denims, underwear, casual wear clothing and towels can be made from lyocell fabrics. The fabric has silkier appearance and hence it is a very good material for ladies wear and shirts. It can be blended with cotton, polyester, linen and wool to impart special properties to the fabrics. Conveyor belts, specialty papers and medical dressings are also made from lyocell. Like Tencel, it is popularly used for making diaper-wipes.

1.17.2   POLYAMIDE FIBRES

Nylon: It is synthetic, thermoplastic and protein-based fibre. The regular fibre possesses round cross-section. The normal nylon fibre is smooth, shiny and transparent. Sometimes, to delustre the fibre, certain pigments are introduced during extrusion through spinnerets. In some special cases, nylon is produced with trilobal or even multi-lobal cross-sections. The production of nylon can be in the form of mono or multi-filament and staple or tow. The fibre during spinning is partially drawn. The fibre has long and straight molecular chain with no side linkage. When drawn, there is better orientation of molecules along the length of the fibre. This makes nylon more crystalline and stronger.

During the manufacture, a product 'hexamethylene-diamine' is formed. It has acid at one end and amine at the other end. It is then polymerized to form a chain. Basically, there are two types of nylons – nylon 66 and nylon 6. The properties of nylon 66 include features such as capability to pleat and crease when heat-set at high temperature. Its melting point is 256°C and is resistant to sunlight and weathering effects. It possesses excellent colour fastness and abrasion resistance.

As against this, nylon 6 dyes easily but fades. It has high impact resistance, greater elastic recovery and rapid moisture absorbency. It exhibits better resistance to fungi, molds, mildews and many chemicals. It is a comparatively stronger fibre and when flamed, it melts instead of burning. Depending upon the drawing, its amorphous and crystalline region proportion can be controlled.

Nylon can be used as the matrix in composites. When these composites are reinforced with glass or carbon fibres, the structure becomes very useful materials in car components around the engine. Its high strength and excellent abrasion resistance makes it most suitable for making ropes and various types of nets, including fishing nets, parachute fabrics and tyre cords. In early days, nylon saris, especially Twinkle nylon sarees were very popular.

But such sarees were found to be hazardous when working in the kitchen; as they easily caught fire. The crinkle nylon (supposed to be accidental discovery) is used in hosiery garments. It is one of the best materials for brush bristles, and hence various types of brushes are made from nylon. The straps and belts are very common popular products of nylon. All waterproof clothes like rain-coats, hats, leggings, and veils use nylon.

1.17.3 POLYESTER

It has 'ester' as functional group in the molecular chain. It is actually polyethylene terephthalate (PET) and as such, this synthetic fibre is not bio-degradable. Most of the polyesters are thermoplastic in nature, though they can be also thermo-set. Some of the popular abbreviations are as follows:

PET – Polyethylene Terephthalate; PSF – Polyester Staple Fibre; POY – Partially Oriented Yarn (filament); DTY – Draw Texturized Yarn; FDY – Fully Drawn Yarn; A-PET – Amorphous Polyethylene Terephthalate Film and BO-PET – Biaxial Oriented Polyethylene Terephthalate Film; DMT – Dimethyl Terephthalate

Polyester is manufactured from constituent acids and alcohols which are derived from petroleum. When acid and alcohol are made to react in vacuum at high temperature, there is condensation polymerization. The material is extruded in the form of ribbon from which polyester chips are made. After complete drying, these chip are then heated and melt-spun by passing the molten liquid through fine spinnerets. The emerging fine filaments are cooled by air and wound on suitable package. The filaments thus formed are hot stretched almost 5–6 times (drawing) to the required fineness. These are then wound on cones. Very fine filaments are crimped and cut into staple fibre.

Polyester is a bright fibre when manufactured. But, it can be made dull in appearance by adding delusterant during melt-spinning process. By changing the shape of the spinneret, the fibres with different cross-sections, such as circular, square and bean-shape, can be produced. This changes the hand of the fibre and the fabric made from them. Nowadays, even hollow fibres are made to suit the purpose.

The filament/fibre during drawing can be stretched to a much higher degree to produce micro-fibres. Normal polyester fibre is smooth and rod-like. However, crimping imparts bulk and adds useful properties like warmth, insulation and moisture transmission (wicking). The fibre is popular both in pure form and blended form.

The fibres are manufactured with a normal or high tenacity (4–7 g/den). It possesses high elongation at break (15–25%); however, it hardly can hold moisture (0.4%). The fibre is lighter than cotton (specific gravity-1.38). Its melting point is around 250°C and possesses very high abrasion resistance. It is quite resistant to acids and alkalis at low temperature, but gets degraded with sulfuric acid at higher temperature. Even strong caustic soda dissolves the fibre around boiling temperatures. The organic solvents and bleaching agents do not have much effect on the fibre. It is quite resistant to mildews and insects. The fibre is dyed with disperse and azoic dyes. It can be dissolved in chlorinated hydrocarbons or phenol at high temperatures.

The fabrics woven from polyester (or its blends) are very popularly used in apparels, furnishings, jackets, hats, bed sheets, blankets and upholstery. Industrially, it is used in tyres as reinforcement, for making conveyor belts, safety belts, coated fabrics and in plastic reinforcement for high-energy absorption. The polyester fibre is used in cushioning and insulating material in pillows and in making comforters (quilts) and upholstery padding.

1.17.4 Acrylic Fibre

Poly-Acrylonitrile is the base polymer from which the fibre is produced. The average molecular weight is around 100,000 with about 1800 monomers. An acrylic fibre contains about 85% of acrylonitrile monomers. In such cases, the co-monomers are vinyl acetate or methyl acrylate. The first fibre was 'Orlon'. The other popular trade names are 'Dralon' and 'Acrilan'.

The acrylonitrile monomer is polymerized in aqueous suspension. The dope is filtered and the fibre is produced by dissolving the polymer in either dimethylformamide or aqueous sodium thiocyanate. The solution is metered while passing through spinnerets. The resultant filaments are coagulated by passing them through the same solvent (wet spinning) or by evaporating the solvent in the stream of inert gas (called dry spinning). The filaments are washed, stretched (drawing) and crimped. The normal fibres vary in their fineness (2–15 d).

In appearance, feel, bulk and handle, acrylic fibre resembles wool. For finer and delicate use, very fine fibres (0.6–0.8 d) are sometimes made. The fibre exhibits some very useful properties such as heat retention, shape retention, high durability, quick drying qualities and fastness to sunlight. As the fibre closely resembles wool, it is again popular in knitwear trade. It is sometimes mixed with wool to reduce the cost of mixing. The wool, in this case, does not felt.

There are certain other favourable properties. The fibre shows water absorption and quick water transportation. It also shows resistance to light and weather. It is quite resistance to mineral oil and dilute acids. However, it is only moderately resistant to alkali. It degrades quickly with concentrated acids, alcohol, esters, ketones, vegetable oil and oxidising agents. It is easily attacked by various organic solvents. The fibres are commonly dyed with basic or acid dyes. With a very low moisture regain (1 to 1.5%), there are, however typical problems of static and pilling.

Acrylic fibre is lightweight and soft. It is a cheaper alternative to cashmere wool. It is used for sweaters, rugs, mufflers, socks, shawls and other winter wear. It is also used as pre-cursor (predecessor) for carbon fibre. Acrylic sheets are used for making beautiful paintings. Acrylic mats are cheaper alternatives. Adhesive acrylic tapes are useful additions. The wrinkle-resistant fabrics are best made with acrylic fibres.

1.17.5 POLYPROPYLENE FIBRES (PP)

When petroleum is cracked catalytically or even thermally, one of the constituents obtained is Propylene. After the process of polymerization, maintaining certain conditions of temperature, pressure and time, polypropylene is obtained. For this propylene is dissolved in suitable solvent (heptane – C_7H_{16}) using catalyst. Under very pressure (around 100 ata), the conditions are maintained for 8 hours at around 100°C.

The polymer has an average molecular weight of about 250,000 [depends upon number of repeat units (C_7H_{16})]. The polypropylene is melt-spun. The filaments are extruded at 220°C–240°C (much above the melting point), cooled in air chamber and collected on bobbins. The filaments are hot drawn (polyethene – cold drawn) and twisted into yarns.

PP fibre is colourless and has smooth surface. It is circular in cross-section. The fibre tenacity varies from 4.5–6.0 gpd, whereas elongation at break is quite high (17–20%). It does not hold moisture (moisture regain almost zero). When boiled in water for about 18–20 minutes, it shrinks (15–20%). It is lighter than water (Sp. gravity 0.92). The fibre softens at 150°C and melts between 160°C and 170°C. PP has excellent resistance to organic solvents, though, in presence of sunlight, the oxygen attacks the fibre. The common chemicals do not damage the fibres. The most important and useful property is that it shows sustained resistance to insects and micro-organisms.

There are many applications of polypropylene fibre. It can be used in making tiles (vinyl tiles) or making curtain poles, especially for babies. In fibrous form, it can be spun into ropes or cords and woven into upholstery fabrics, carpets and filtration materials. It can be used to make horticulture or agriculture components, medical care supplies and disposable nappies. It can be used as chemical binder in non-woven material when softened.

Being environmental friendly, artificial grass lawns are made. In the apparel form, it is used as thermal underwear and lining. It is very popular in the automobile industry for making side doors, panels, armsets, interior lining etc. The wardrobe doors of PP are very popular. It is used in drainage pipes and woven/knitted tapes. As it strongly resists microorganism attacks, it is extensively used in 'Geo-Textile' (canal linings, soil erosion, as reinforcement in concrete – crack-preventing and road building).

1.17.6 ARAMID FIBRE

They are altogether different classes of fibres and basically are 'aromatic poly-amide'. They are very strong and heat-resistant synthetic fibres. The very high strength is owing to their highly oriented molecular chain along the fibre axis. The first fibre in the series has been called Nomex. It possesses excellent heat resistance and yet has almost identical handle like other textile fibres. Under normal atmospheric oxygen level, it does burn or ignites. The fibre is also manufactured under another trade name – 'Tejinconex & Kermel (France). Later, 'Kevlar', a fibre with a much higher tenacity and elastic modulus was developed. These fibres are mostly used for high-tech applications. There is another brand name under the same category – 'Technora'.

It is the reaction between amine group and halide group that leads to aramids. The polymer is dissolved in anhydrous sulfuric acid and is subsequently produced with the help of liquid chemical blend. The fibre shows strong resistance to abrasion, organic solvents, heat and flame. Even at 500°C, the fibre does not easily degrade. Owing to almost no moisture content, it easily builds up static. Also, the fibre shows sensitiveness to acids and ultra-violet radiations. Kevlar has very high strength, high Young's modulus, high tenacity, low creep and low elongation at break. But its dyeing is difficult. Hence, solution dyeing is used.

There are many industrial applications of aramid fibres. As it is flame retardant, flame-resistant clothings, heat-protective clothings, hot air filtration fabrics for helmets and body armors are made from the fibre. In present days, therefore, it has almost replaced asbestos. It is a very useful material in composites. It is very popularly used in making tyres, mechanical rubber-goods, ropes, cables and optical fibre-cable systems. The variety of uses still continue when sailcloth, sporting goods, reeds for wind instruments and loudspeaker diaphragms are made from the fibre. The sports material such as tennis strings, hockey sticks and snowboards are usefully made from the fibre.

1.17.7   POLYETHYLENE

The monomer in this case is ethylene, and it is mainly held within carbon-hydrogen chain. After polymerization a really long-chain molecule – polyethylene is formed. It is a thermoplastic polymer with chemical formula $(C_2H_4)_n$. The density of fibre ranges from 0.92 to 0.94 g/cm^3 (a medium density fibre) and from 0.88 to 0.91 g/cm^3 (a low density fibre).

Both the melting point and glass transition temperature vary according to the crystallinity and molecular weight (for high density – around 125°C); for low density – around 110°C).

Both high or low density grades have strong resistance to chemicals and hence do not dissolve at room temperature. This is because of its higher crystallinity. The fibre/film shows very good resistance to strong acids and base. It is also resistant to gentle oxidizing and reducing agents. While burning, there is paraffin odour and the blue flame has slight yellowing at its tip. The polyethylene can be dissolved at higher temperatures in toluene and xylene.

The ethylene molecule is very stable, and it is polymerized using catalyst titanium chloride. The polyethylene is not bio-degradable except when exposed to ultra-violet light. This leads to many environmental problems. Recently, it has been discovered that some types of bacteria can degrade polyethylene up to 40% of its weight, and that too after a really long time. The polyethylene with ultra high molecular weight has a very tough structure and hence it is used in making bottle handling machine parts, moving parts on weaving machine, bearings, gears and butcher's chopping boards.

Due to their stronger intermolecular force and strength, high density polyethylene can be used in making milk jugs, detergent bottles, garbage containers and water pipes. Medium density polyethylene is a good shock absorber and possesses drop resistance. Hence, it can be used in making gas pipes and fittings. Sacks, shrink-films, packaging films and carry bags can also be usefully made from polyethylene. Owing to lower intermolecular force, lower tensile strength and increased ductility; the low density polyethylene is mostly used in making plastic films and bags-wraps.

1.17.8 ELASTOMERS [ELASTIC POLYMERS]

It is a polymer with 'visco-elastic' property. It means that they have very low Young's modulus and high yield strain as compared with other fibres. Each monomer is usually made of carbon, hydrogen, oxygen and/or silicon. They are amorphous in nature. When the stress is removed, they spring back to their original configuration. Thus, they show very high flexibility, even up to 700% stretch. When cooled to crystalline phase, the chains become less mobile and consequently show less elasticity, than the one manipulated, at a temperature higher than glass transition temperatures.

1. **Spandex [or Lycra, Elastane, Elaspan, Linel]:** All of them possess very high elasticity, the word spandex being an anagram of another word 'expand'. They are more strong and durable than ordinary rubber. It is a co-polymer made of polyurethane and polyurea.

Different methods are used in making these fibres. They are extruded through spinneret, reaction spinning, solution dry spinning or wet spinning. Before following any of these methods, a long-chain pre-polymer is first made. The polymer is then reacted by choosing any of the above methods to form a long continuous fibre. Even then, the solution dry method is a little more popular. The spandex filaments are generally extruded through circular spinnerets. However, during drying, they deviate a little from this shape. In multi-filament yarns coming through spinnerets, the individual filaments are very likely to get fused at places. Usually, the number of filaments in a yarn varies from 12 to 50. With their linear density varying from as low as 0.3–3 tex. The density of the spandex (filaments) is 1.15–1.32 g/cc (a value lower than cotton). When surface finish is removed, the moisture content in the fibre varies from 0.8 to 1.2%. From the point of view of fibre strength, they are very weak (0.6–0.9 g/den). The fibre is made by producing a fine filament and then cutting them. Both, the fibre and filaments are a little whitish. They show a dull lustre and possess outstanding elasticity. The fibre slowly burns at around 150°C and yet shows heat resistance with varying degradations. It is low electrical conductor.

Unless exposed for a long time, it exhibits good resistance to most of the acids and is moderately resistant to alkalis. It offers good resistance to dry-cleaning solvents. It dyes in a full range of colours. Further, during bleaching, chlorine-based bleach should be avoided as it degrades the fibre.

These fibres or filaments are extensively used in best-fit fabrics, where high stretch is combined with comfort in clothings. These criteria are very much required in various sports and athletics activities such as aerobics, exercises, bathing/swimming/wet suits, body-fit garments, elastic strips, disco jeans, slacks, hosiery, socks, diapers, underwears, dance belts, cycling shorts, support hose, surgical hose, motion capture suits and shaped garments.

In clothing, it is used in very small percentage. It is usually mixed with some other textile material – cotton or polyester to reduce reflection of light.

1.17.9 POLYURETHANE (PUR OR PU)

A polymer has a chain of organic units, joined by urethane. The 'step-growth' process is used to convert monomer into polymer. Two monomers one containing at least two iso-cynate functional groups and another monomer containing at least two hydroxyl (alcohol) groups are used in presence of a catalyst. The characteristics of polymer are improved by adding surfactants, flame retardants, and light stabilizers.

Fully reacted Polyurethane polymer is chemically inert. It is combustible solid. When burnt, it releases harmful gases. Its dust can cause irritation to eyes and lungs. For this, respiratory gadgets like dusk mask are required to be used when tackling fire problems.

It is a unique material offering elastic rubber with toughness and durability resembling metal. It is available in varying hardness from soft rubber to bowling ball. It has almost replaced rubber, plastic and metal in many applications. It shows very good abrasion and tear resistance and gives load-bearing capacity. It has a much better impact and wear-resistance and shows a good elastic memory. Owing to excellent noise reduction property, it is used where the noise level with machine part is very high. Its electrical insulation property is good. It can withstand heat up to 80°C. However, beyond this temperature, there is degradation. The polymer shows good resistance to oil, solvents, grease and gasoline.

Polyurethane products cover extremely wide range varying in stiffness, hardness and density. Urethane has replaced metals in sleeve bearings, wear plates, sprockets, rollers and various other parts. When used in machine parts, it gives benefits of weight and noise reduction and decreasing wearing effects. It is used in cast sheets and fabric-backed sheets, as it can be moulded to précised size and shapes. It can also be used in various types of pads, e.g. tyre pads, wear pads, gaskets and dumping pads. Polyurethane is the best material for flexible and rigid foam seating. Also, it finds its way in carpet underlays and foot-wears.

Polyurethane when used in paints gives best protection to wooden furniture. The sealants of polyurethane are used to prevent air and water leakage. The boat hulls are made of rigid polyurethane foam core, as it is very strong and yet buoyant. Tennis over-grips are made from fibre material. A thin film coating of polyurethane is used for lamination purpose. The normal fibre is capable of being stretched to as high as 600% and it can still come back. The fibre is sometimes quoted with nickel or silver to protect it from static or from bacteria attack, respectively. Especially, the coating can be used to radar chaffing.

1.17.10 CARBON FIBRE

These fibres have higher stiffness and tensile strength. They show greater resistivity to temperature and thermal expansion. Their low weight and chemical resistance makes them useful in aerospace technology and for military purpose. However, as compared to the fibres like glass showing similar properties, they are costlier.

The raw material for making the fibres is called 'pre-cursor'. The raw material used in a majority of the cases is PAN (polyacrilonitrile), which is a long-chain polymer. Few other manufacturers use rayon. The process of conversion is partly chemical and partly mechanical. The raw material (pre-cursor) is drawn and heated to very high temperature in the absence of oxygen. The carbon atoms while vibrating violently drive away the other non-carbon atoms – 'carbonization'. The fibre, thus remaining, contains long-chain carbon atoms which are very tightly bound. The fibre surface is slightly oxidized. The surface treatment is given. The fibres, at this stage, needs some protection and hence sizing is done. The coated fibres are wound and finally twisted.

The fibres are usually used in composite structure which are very light in weight and yet have very high strength. Since the last decade, the demand for carbon fibres has rapidly increased – especially in the field of aerospace and air craft. The fibre is used in high temperature applications and possesses excellent corrosion resistant. The fibre can be used to make fabrics, micro-electrode, flexible heating elements and few car parts[2,5].

REFERENCES

1. Introduction to Study of Spinning – W.E.Morton & G.R.Wray.
2. Manual of Cotton Spinning – "Characteristics of Raw Cotton", Vol. II Part 1, Textile Institute, Manchester, 1961.
3. Wikipedia.
4. Introduction to Textile Fibres - H.V.Sreenivasa Murthy, Textile Association Pub.
5. Textile Science, E.P.G. Gohl.
6. Textile Fibres, V.A. Shenoi.
7. Modern Fibres – Woodhead Publishing Ltd. – Tatasuya Hongu & Glyn O. Phillips.

2 Cotton and Its Cultivation

2.1 GENERAL INFORMATION

Cotton falls into the category of seed hair, the only other fibre of this type being Kapok. However, unlike cotton, Kapok is very short length fibre and quite weak.

Cotton is a soft, fluffy, staple fibre that grows on seeds which are enclosed in a form known as a boll. The cotton plant looks like a shrub popularly grown in tropical and subtropical regions around the world. Different countries - U.S.A., Peru, Brazil, India, Africa etc. and now China grow plenty of cotton. The fibre is most often spun into yarn or thread and is used to make a soft, breathable textile. The cotton is most widely used natural fibre in clothing today.

In the beginning of 20th century, a wide spread disease in Peru spread over the country. It went through roots and stems of the plant and totally dried them. With lot of experimentation and research, Herman Tanguis[2] was able to develop a far better variety which was resistant to the disease. The plant was produced from the seeds. It grew 40% longer fibre (between 29 mm and 33 mm) and yielded stronger fibres which did not break easily. The plant also required much less water. This variety became very popular and almost 75% of it has been used for domestic market and apparel exports.

Another large exporter of raw cotton is Africa where the share of the cotton trade has doubled since 1980. Neither U.S.A. nor Africa has a significant domestic textile industry. This is because the textile manufacturing has been moved to developing nations in Eastern and Southern Asia such as India and China. In Africa, there are many small cotton growers producing two very popular and leading varieties - Memphis and Tennessee cottons. Accordingly, there are good numbers of ginneries in many African states.

The top six cotton producers in the world are: China, India, U.S.A., Pakistan, Egypt and Uzbekistan. Likewise, the top six cotton exporters are: U.S.A., India, Uzbekistan, Brazil, Australia and China. Equally interesting will be the fact that there are few countries that are non-producing cotton but still heavily import cotton e.g. Korea, Taiwan, Japan and Hong Kong.

In the the United States, Texas leads in total production while the state of California has very high yield per acre. In India, the leading states producing cotton are - Maharashtra, Gujrat, Andhra Pradesh and Madhya Pradesh. All of them have a predominant tropical wet and dry climate. Even then, with agricultural research being encouraged all over India, and the knowledge being passed on to the farmers to improve both quality and quantity, the scenario is steadily changing.

Photograph taken by the author in California, Cupertino

FIGURE 2.1   Opened Cotton Bolls – When the bolls burst open, the cotton is almost ready to be picked-up.

2.2   OVERALL COTTON CULTIVATION

Successful cultivation of cotton requires a long frost-free period, plenty of sunshine, and a moderate rainfall, usually from 600 to 1200 mm (24 to 48 inches). The soil usually needs to be fairly heavy, although the level of nutrients does not need to be exceptional. In general, these conditions are met within the seasonally dry tropics and subtropics in the Northern and Southern hemispheres. But a large proportion of the cotton grown today is cultivated in areas with less rainfall and that the water obtained is from irrigation. The production of the crop for a given year usually starts soon after harvesting in the preceding season. The planting of the seeds, in northern hemisphere varies from February to June. The south plains in U.S.A. are very large cotton-growing region. The cotton in dry land is successfully grown with irrigation water.

In India, the cotton in many parts of the country is a monsoon crop when the yield depends largely on the vagaries of the nature. However, the cotton grown on irrigation is very large in the world. This is because the best growing conditions can only be obtained in this situation. In the countries where the severe frost problems arise, the cotton growing can seriously get affected both in quality and quantity. It is known that the pests and insects grow rapidly when the atmospheric conditions like moisture, shades and relatively cool temperatures prevail. Though frost is an enemy for certain types of insects and pest, the growing season, in such situations is required to be prematurely closed.

In India, the time of cultivation of cotton varies. Even the method of cultivation and the types of cottons grown are choices depending upon the local practices. Immediately after the independence, the grades of cotton were poor. However, then onwards, far superior qualities are grown, many of them being hybrid types. Shankar-4, Varalaxmi, Suvin are some of the popular higher class varieties. Even then, due to poor picking methods, the percentage of dirt, leaves and trash etc. is comparatively higher, thus making their processing problematic.

The seed distribution and multiplication are the main issues in deciding the quality of cotton grown. But the relative cost involved and the expected yield that a farmer can expect, do not show a good correlation, owing to poor practices of ploughing and non-assurance of regular and timely water supply. The small farmers cannot afford ploughing by machines, which again affects the quality. Many a time, the farmers have to depend upon the first rains before they can sow the seeds. In many other instances, after the seeds are sown, there is neither adequate water supply through rains nor any provision for irrigation or well-water supply during early growth. When bullocks are used to plough the fields, it is not possible to go deep into the land. The sowing period also varies across the country; for example in Maharashtra, it is done in the beginning of monsoon in the month of June, whereas in south, it is further prolonged. This again is largely governed by the rain water.

While sowing the seeds, the spacing between the two adjacent plants is equally important. This ensures adequate nourishment for them. The thinning out operation is carried out in about two months after sowing. The 'Hoeing' operation again is very important to remove any extra, unwanted growths around the main plant. Even during the growth, the land around each plant needs some preparation in the form of loosening the soil, so that the water goes into the soil and does not get evaporated.

The flowering starts after about 4 months. The flowers are initially pinkish-white and tiny. This is the period for fertilization, after which the flowers wither out and form small capsules. This is when the height of the plant is approximately one meter. The extra leaves on the lower side of the plant are purposely removed so as to provide maximum nourishment to the capsule or pod and the seeds inside. This helps in providing better nourishment to the seeds and the hairs which continue growing on it, thus improving the cotton fibre quality.

As the growth continues the seeds are covered with numerous, tiny fibres. There are, on and average, 4000–4500 fibres on each seed. The fibre growth is from inside of the fibres to the outside towards the walls. All this, during the growth, is packed into a tiny pod or "Boll" which with the growing mass also swells. When the mass can no longer be accommodated within the bolls, they burst open. It is also an indication that the growth is nearly complete, some fibres still continue to grow though.

The bursting open of the bolls exposes the white fluffy mass of cotton. A great care is required to be taken after this. It may be even necessary to spray pesticides on the open masses so as to protect them from any mildew attack. The exposed mass is usually allowed to remain under bright sunshine so as to dry the wetness inside the bolls. It also allows reaching of some more nourishment to those fibres which still have not fully grown. However, in the case of cloudy and shady atmosphere, the bacteria grow rapidly and it is advisable to decide upon an early picking date.

In many parts of country in India, labourers are employed to do the job of picking. In other countries, especially where huge fields are used for cotton growing, the machine-picking is resorted to. Whichever is the picking method, one thing must be ascertained that the cotton mass picked-up must be fully dried. Though in some instances wet cotton is picked up to increase the weight of seed cotton, during its subsequent storage, it is susceptible to attack from mildews and bacteria. The wet cotton also makes the job of ginning more difficult. Though the cost of picking is more in the case of hand picking, the employed labourers when trained, do the job more carefully, thus selecting only ripened and fully dried cotton. Especially when the cotton of high quality is required to be picked-up, this type of care during picking helps quite a lot in avoiding association of any vegetable originated impurities.

Elsewhere, in the world, especially in Russia, where rain fall is scanty, evaporation of water becomes a great problem. The fields are usually laid out in ridges and furrows for better placement of sowing of seeds and then plants. When the soil is hard, many seeds are sown in one hole so as to enable them to easily break through. Only a limited number of watering is given so that the plant just gets adequate supply of water. Too much of watering must also be avoided; otherwise the soil becomes harder and the plant growth gets affected. The transverse furrows perpendicular to each other are made in such a way that the plant is placed at their center, forming a sort of island. This enables irrigation water to reach it without much of wastage. When the surrounding area is very hot, the sweeping hot air currents and the high winds blowing the dirt and trash over the opened out bolls add contaminants to the fluffy mass of cotton.

In U.S.A., cotton is grown on both rain and irrigation water. However, the climate is very uncertain. The rains during spring are sometimes heavy and make the soil hard. Heavy floods of the river (Mississippi) sometimes pose problems. Summer crop is at mercy of the weather, too much precipitation or event of drought both being harmful. The first killing frost restricts the growth of the plant and cotton maturation period gets extended. Sometimes it is blessing in disguise as it also destroys the boll-weevil. Further, the date of sowing and picking are also variable factors. The sowing is delayed till the frost season is safely over. Mechanical means are provided for ploughing. The cultivation depends on use of fertilizers when the crop is to be repeated. The sowing machines are used which finish the vast job quickly. After thinning or chopping, the plants are spaced at a distance of approximately 1–2 feet. As usual, the weeding operation removes any unwanted growth around the main plant. The ground surrounding the plant is also broken at intervals to avoid quick evaporation of water.

The flowering starts in 2–3 months after sowing when the plant reaches a height of about 2 feet. The flowers change their colour from light yellow to pinkish purple. Subsequently they fade and wither out. The bolls are formed in about 3–3½ months. The picking, in the favourable atmosphere is extended and there are sometimes, as many as three pickings. After picking and before ginning, sometimes seed cotton is stored for certain duration and it is observed that the fibres continue their growth from seed nutrition. The cotton when picked early and is still wet, it also allows the time for full drying. This helps in improving fibre uniformity.

In Egypt, most of the very superior cotton varieties are grown around the banks of river Nile. There is hardly any rainfall and hence most of the cultivation is in the narrow belt around Nile which itself has no tributary. This restricts the cultivation on a strip of about 550 km with hardly 1 km width on either side of the banks. The crop thus grown has to solely depend upon the irrigation water, though occasionally there are few wet days in winter.

The term Egyptian cotton refers to the extra-long staple cotton grown in Egypt. It is mostly used for luxury clothings and by top market brands worldwide. Egyptian Giza cotton is more durable and softer than American Pima cotton, which is why it is more expensive.

The ground is ploughed 3-4 times, every time at right angles to previous one. The ridges are ploughed and shaped. The time lag of nearly one month between ploughing and sowing is considered advisable. The seeds are sown in dry condition after which watering follows. The protection from the north wind is sought while sowing the seeds on south side of the ridge. In thinning out, not more than two plants are allowed in one hole. There is no killing frost and hence weather hazards are quite low. The sowing is planned in such a way as to avoid danger of boll worms in autumn. Hoeing is followed after 3-4 weeks to remove weeds and break the soil. If necessary, the operation is repeated. During all this, timely watering is carried out. The number of pickings can be more than one. However, the maximum bulk is picked-up in the first picking itself. The sowing seeds for the next cultivation are mostly recovered from this first picking.

The cultivation of Sea Island cotton is mostly in the region of West Indies islands and only a small quantity is grown in Florida, Georgia and South Carolina. The original variety is however, "Gossipium Barbadense", which is grown in West Indies islands. The quantity grown is, however, relatively small and it needs the people of great experience to handle the cultivation. The production of this variety varies greatly owing to the vagaries of nature, especially the hurricanes around the island, and hence very often both the quantity available and the price fluctuate. Further, due to mild and humid conditions and also mitigation of winter frost, the ambient conditions are more favourable to boll weevil attack. This gets further aggravated by late maturing period required for this variety.

St. Vincent is one of the best Sea Island varieties grown in West Indies. Even then, the short of labour and method of cultivation lead to low yield. Further the soil is volcanic and the soil surface is mostly rocky, thus making mechanical ploughing more difficult. The cotton being the main crop, there is no catch-crop in between the two cotton crops. The plant being tender, it is more susceptible to pest attack. The fibre is very delicate and sometimes machine pressing is avoided to prevent any damage to the fibre. The seeds after ginning are consumed locally for feeding stock, obviously without extracting oil. When oil is extracted, it is mostly used for culinary purpose.

Another variety of cotton is Shiny cotton. It is a processed version of the fibre that can be made into clothes resembling satin for shirts and suits. However, its hydrophobic property of not easily taking up water makes it unfit for the purpose of bath and dish towels.

2.3   AREA & YIELD

Cotton, often referred to as "White gold", has been in cultivation in India for more than five thousand years. Though synthetic/man-made fibres have made in-roads in many countries in the world, cotton holds the prime position in India, constituting more than 70% of the total fibre consumption in the textile sector.

In India, cotton is grown over an area of about 9 million hectares and provides livelihood for over 4 million farming families. Yet, in regard to productivity of cotton, we are far behind the other cotton producing countries. This is because, in India, more than 70–75% of the cotton is cultivated without suitable irrigation facilities and it solely depends on rain. In addition, all the farmers are not able to use good seeds and manures.

While per hectare yield of cotton in India is as low as around 330 kg, a small country like Turkey produces almost three times this quantity per hectare and occupies very high rank in the world. Per hectare yield of cotton in U.S.A. is around 680 kg and in China it is around 1000 kg. The higher productivity in these countries is mainly due to innovative and modernized method of cultivations.

The economic reforms and the trade policy liberalization carried out during the last decade with a view to globalizing the Indian economy have exposed the Indian cotton textile industry to a new challenge. In the last two decades, the production of cotton has gone up significantly (around 470 kg/hectare). This is basically due to the introduction of high yielding and hybrids varieties. In addition, improved methods of management of insects, pests and diseases have been introduced.

2.4   COLOURED COTTON[8]

Naturally coloured cotton is the one that is grown to have other than the its lemon-yellow or off-white colour. The colours are usually red, green and several shades of brown. This is a natural colour and does not fade with the time. The yield is, however lower and the fibres are shorter in length. It is also a weaker fibre, but has a softer feel. The coloured cotton is not commonly grown as it requires special agricultural techniques. So also, it is more expensive to harvest these cotton varieties. Therefore, many countries producing this cotton shifted back to cultivating natural white varieties.

The colour that the cotton gets is due to certain pigments e.g. green colour is derived from typical wax layers whereas brown cotton gets its colour from tannin in the lumen of the fibre. This cotton being too short in staple length was thought to be unsuitable for normal clothing manufacturing. The efforts however have been made to make the fibre longer and stronger by continuous research work. With the improvement in the technology, it is now possible to get four different colours – green, brown, red and mocha (tan).

All these cottons, especially those with the green colour have excellent sun protection ability. It is thought that the pigments within the fibre have potential to shield ultra violet rays (UV-protection). Being coloured, it does not require dyeing. Naturally coloured cotton opposes the change and hence, the colour becomes more intense after each washing and laundering. When exposed to bright sun light, however, the colour is susceptible to change. It is observed that the brown cotton is more stable. The 2.5% span length varies from 20–24 mm and fibre strength varies from 16–19 g/Tex. There is greater variation in micronaire – with a range from minimum 3.2 to 5.4 (maximum). There are certain benefits of using coloured cotton:

TABLE 2.1
Fibre Properties of Some Coloured Cottons[6,7]

Variety	Yield g/ plant	2.5% Span Length-mm	Mc	Maturity Coefficient	Fibre Strength – '0' gauge
Vikram	207	20.7	3.1	0.76	37.1
MCU-5	195	23.3	2.7	0.62	36.3
CNH-36 Parbhani	169	19.3	3.1	0.76	37.8
LRK-516	130	24.4	2.5	0.57	37.2
LRA	173	25.7	3.4	0.77	45.3

The criticisms are always laid that the artificial dyes may lead to allergies, itching or even cause skin cancer. Using coloured cotton has obviated these adverse ill-effects. With the use of naturally coloured cottons, there is great saving in cost, which otherwise is quite high. The environmental pollution due to chemical residue or affluent in finishing processes is fully absent when using coloured cotton.

However, there are certain drawbacks. The fibre yield during cultivation is much lower. The varieties show poor fibre properties. With artificial dyes, sky is the limit for colours and their shades. This is not possible with naturally coloured cotton. Owing to less demands, the proper marketing facilities are not still available. During cultivation also there is a fear of contamination of colours. The cotton being cross pollinated crop, the growing of coloured and white cotton side-by-side would often lead to colour-contamination. This contamination can still occur even at ginning and in subsequent de-linting process. The fibre properties of some known coloured-cotton varieties are given in Table 2.1.

2.5 GENETIC MODIFICATION[2.4]

Genetically Modified (GM) cotton was developed to reduce the heavy reliance on pesticides. The bacterium Bacillus Thuringiensis (BT) naturally produces a chemical harmful only to a small fraction of insects, most notably the larvae of moths and butterflies, beetles and flies, and is harmless to other forms of life. The gene coding for BT toxin has been inserted into cotton, causing cotton to produce this natural insecticide in its tissues. In many regions the main pests in commercial cotton are lepidopteron larvae, which are killed by the BT protein in the transgenic cotton that they eat. This eliminates the need to use large amounts of broad-spectrum insecticides to kill lepidopteron pests. However, some of the pests have developed parathyroid resistance.

BT cotton is ineffective against many cotton pests. However, depending on circumstances, it may still be desirable to use insecticides against these. A study carried-out on BT cotton-farming in China revealed that the presence of other pests necessitated the use of additional pesticides and this level was almost at par with that required for non-BT cotton, thus causing less profit for farmers. It may also be noted that, as it is, GM cotton seeds are quite expensive. It was reported that GM cotton was planted on an area about 20% of the worldwide total area for cotton in 2000. The U.S. uses about 70% of its cotton planted area for GM varieties.

The initial introduction of GM cotton proved to be a huge success. Though the yields were equivalent to the non transgenic varieties, much less pesticide was used to produce cotton. Subsequent, the introduction of a second variety of GM cotton led to increases in GM cotton production. The cotton has also been genetically modified for resistance to glyphosate, an inexpensive but highly effective broad-spectrum herbicide.

GM cotton acreage in India continues to grow at a rapid rate. With about 9.0 million hectares, it makes India one of the countries with very large area of GM cotton under the cultivation. As claimed by the suppliers of Bt. Seeds, the major reasons for this increase are a combination of increased farm income and a reduction in pesticide use to control the Cotton Bollworm.

2.5.1 OTHER SIDE OF BT COTTON

Initially, it was believed that the transgenic cotton would pollinate other plants in the nearby area. The trials taken in this respect showed that in very few cases, the pollens from BT cotton really travelled over a great distance. Further, it was also observed that the pollens from BT were incapable of pollinating with any other plants but cotton. It was found that resistance against the boll worm was especially in-built in Bt cotton. The proteins injected in the plant for this purpose do not cause any harm to other beneficial insect species. Even it was found that BT was quite safe to birds and mammals. Though BT was supposed to be a terminator gene (it is not carried to the next generation seeds), there were several generations which had moved along (advanced). Even then, there are certain matters that require proper attention. The effect of this gene on ecology needs to be further studied. The points related to this are whether it really spreads and if so, in what conditions. Equally possible is that trade with BT seeds is likely to be in the hands of only few companies which can have hold over its sale, especially over its cost. This is because; it is found that the cost of BT seeds is very high as compared to ordinary cotton seeds. There is always fear that the pests would develop resistance against BT genes, especially in the subsequent generations. In this case, it must also be ascertained that whether any additional spraying of pesticides would be necessary. Further, the cotton seed cakes are the best food for animals. Even there are some other uses of seeds in food industry. Against this background, it must be also ascertained whether BT seeds are harmful. Finally, it has to be observed, if there are destroying effects on the other native varieties grown around the BT plantation; otherwise it would seriously affect bio-diversity.

There was a typical case where, the local cotton was transformed and grown with BT-seeds. As there was no approval from concerned genetic authorities, the seed and the crop were considered unlawful. As against this, the Govt. had already given official permission to only one company from India having a partnership with a multinational company from outside India. In all, three varieties under this partnership were given permission for transformation to BT. However, majority of the

Indian farmers stuck to locally transformed cotton. This was mainly because; the farmers experienced much higher yield from this locally transformed BT. Seeds. In addition, the cost of the seed packets from this locally transformed Bt cotton was made available at a considerably low price. As for the farmers, no assurance was given for the performance of official BT varieties or even an improved yield. Actually, the locally transformed BT variety led to earlier crop and this helped the farmers to take double crop in a season.

In spite of all these points in favour of locally transformed BT cotton variety, a drastic and unjust decision was taken to remove and burn the whole crop which was fully ready for picking in that season in India. The reason given for this extreme decision was that the locally transformed variety had a very harmful gene. In reality, the local seed company has used a popular cotton variety of the area for transforming it into BT type. Before selecting this cotton, some research regarding suitability, sustainability, its performance in terms of yield and adaptability were taken into consideration. Most surprising was that, later Govt. also tested environmental issues, if any, of this gene. When good and stable performance was seen, the Govt. did allow a limited distribution of this locally transformed BT variety. A comparative study carried out[3,4] for various varieties then grown in the State is given in Table 2.2.

During that particular year, bollworms attacked very heavily on all cotton crops. Most surprising thing was that only locally transformed BT variety sustained and survived. Many other high yielding hybrids were greatly affected by the boll worm attack. Thus, a locally transformed variety showed an outstanding performance. The news spread and all the farmer-cultivators were enthused. However, this upset the authorised Indian company partnering the multinational company. They feared to lose the hold on Indian seed market. But the farmers had tested the fruits of success with locally transformed cotton-gene. They therefore preferred the locally transformed cotton seeds which were quite cheap. Multiplication of the seeds of this cotton continued and grew in large quantity. The area under its cultivation also increased. Even then the issue lingered on a single point – permission from

TABLE 2.2
Comparative Study of BT & Non-BT Cottons[4,7]

Character	50% Flowers ready - Days	Avg. Bolls on plant	Yield – Kg/acre
A – BT Cotton	76.96	95.35	1229.51
A – Normal Cotton	95.39	66.65	800.73
B – BT Cotton	75.95	89.80	1327.83
B – Normal Cotton	88.92	70.89	936.63
Farmer's BT Cros77	75.25	98.62	1198.83
Farmer's Normal	95.32	66.32	775.87

A – Locally transformed gene cotton
B – Genes from Govt. approved company collaborating with multinational

environmental department. It was equally surprising to note that the agricultural department had given green signal to grow this locally transformed cotton-gene seeds. The pros and cons of all such factors needed to be carefully found out.

2.5.2 BT Cotton in India[3,4]

It seems that there were certain shortcomings in implementing programme of BT cotton in India, especially in the initial stages. Normally, pest is a prey for certain birds and other predators. When the pest attack is much less, the predators and birds are not able to fully depend upon the pest as their food. A more detailed study is required on this factor. As mentioned earlier, equally possible is that the pest itself may develop immunity (resistance to pesticides or BT genes). Yet another issue is whether there is a possible diffusion of BT genes in the soil and if there is, would it affect the other useful micro-organisms. It is also possible that the BT gene may develop some allergic effect on seeds. This is very important because, the seeds are popularly used for oil extraction (Lin-seed oil). Lastly, the most important factor is the cost of the BT seed packets i.e. whether the seeds would be available to the farmers at a considerably lower price. A confirmed data should be also available as regards the yield and purity of the seeds. The pros and cons of all these factors need to be carefully studied and publicised.

Two BT varieties introduced by Govt. approved company and two local varieties (non-BT) were compared[3,4] for their performance on the field. It was observed that BT was short duration crop - (90–100 days) as against non-Bt (100–120 days). Both the types (BT & non- BT) gave the boll size ranging from small to large. A very interesting finding from this study was that BT in general, protected the cotton plant from only boll weevil. It did not give any protection to the cotton crop from any other pest (especially from pink boll-worm). The study almost led to an important conclusion that BT crop may not be so profitable, considering, its limited protection coverage, yield, cotton quality and cost of production of the whole crop.

2.5.3 Possible Causes & Their Solutions

As for Indian conditions, it is absolutely necessary to give the farmers proper training. This will give them not only the knowledge of growing BT varieties but

TABLE 2.3

Comparative Performance of BT and Non-BT Cottons[3,4]

Particulars	Non-BT	BT
Avg. No. Bolls/Plant	60–80 % more	50
Boll Size	About 30% more	3.5–5.0 g
Fibre Length	13–15% more	30.5 mm
Cotton Grade	One grade higher	B & C
Quintals/Acre	15–20% higher	2.0–3.5
Quintals/Hectare	15–20% higher	4.94–8.65

also will boost their confidence, especially in sowing methods, watering frequency, additional pesticides, if and when to be used etc. The farmers are often seen to be confused about when to use additional pesticides. These pesticides are most effective only when the there is severe pest attack and also when the level of attack is maximum. The farmers need to be given this training so that their spraying efforts will not be in vain. It is not at all sufficient to merely distribute the seed packets or simply issue some pamphlets. This is because the general education level of farmers in India is not so high. When the cotton or its seeds are required to be stored, there is a need for a strict control. This again is because of hazardous micro-organisms. In the absence of such strict control, it would not only violate environmental safety but also disturb ecological balance.

Owing to very high cost of the BT seeds, it is necessary to ascertain that the cotton varieties chosen to transform them into Bt. should perform well. Any chosen variety or varieties should be suitable for the region where the farmer is going to sow the seeds. Ultimately, all the exercises in developing the BT genes in any cotton variety should bring economical benefit to the farmers. A poor choice would lead to a poor performance of chosen varieties. In the past, this was one of the reasons for the losses that the farmers had to bear. A poor choice often used to lead the farmers to use additional sprays of pesticides. This is very likely to add to their losses. In fact, BT gene does not act on several other types of pests, e.g. pink bollworm, sucking pest etc. The BT toxins have certain period over which they effectively act against the pest. If the period of their effectiveness does not coincide with pest attack, the power of toxins drastically reduces and a full protection from pest is not possible.

It is equally possible that the farmers are easily carried away by the exaggerated advertisements of the seed company, especially regarding prospective higher yield, reduced expenses on pesticides and improvement in some fibre properties. In the case, when the results do not turn out to be as per their expectations, there is only frustration in the minds of the farmers.

Presently, lot of Cotton Research has been directed to careful selection of cottons to be used for BT transformation. In this, the field condition where cotton is grown is also being considered. It is also ascertained that the cotton selected for BT transformation offers best fibre properties and better yield. Earlier this factor was possibly overlooked while selecting a certain variety for Bt transformation.

In this context, BT cotton seed producers in private sector will have to be more alert and take into account the tough competition that they would receive from public sector BT seed companies. This, in the long run, will also benefit the farmers as they are very likely to receive the seed packets at an affordable price.

2.6   ORGANIC COTTON PRODUCTION[5]

The basic principle involved in producing organic cotton is to work with nature rather than against it. Organic Cotton is grown from non-genetically modified plants, without the use of any synthetic agricultural chemicals such as synthetic fertilizers or pesticides. Its production also promotes and enhances bio-diversity and biological cycles. The problem that the organic cotton may possibly cause allergic reactions is most likely a myth and needs to be further investigated. The practices that are to be used while growing cotton are under strict control of an institution - National Organic Program (NOP), U.S.A. The procedures are laid down for pest control, use of fertilizers, and handling of organic crops. In the world, the number of countries growing organic cotton is increasing rapidly. The area under the organic cotton production is also increasing at a noticeable pace. This would increase the organic cotton production sizably in the near future.

There are ways of following certain procedures for cultivating organic cotton. Using compost, recycling the nourishment, rotating crop etc. all help in strictly keeping away the influence of synthetic fertilizers. The soil during this time is made more fertile and productive by using only organic matters. It has been the experience of the farmer around the world that the organic practices in the field have enabled them to save additional expenses on pesticides and synthetic fertilizers. Growing certain favourable plants (like sunflower) around the organic cotton crop also helps. There are certain types of ants that attack and make boll worm their prey. In old days, when sythetic fertilizers were not prevalent in the market, farmers used cattle-dunk which was not only cheap but easily available natural manure. The substances like chili, neem-oil or garlic have the potential to control the pests, and bacterial growths. Their sprays were used on cotton crops to prevent pest attack. These are much cheaper means against pest attack. Sometimes intercrops (pigeon-peas or peas) are taken to also improve fertility of the soil. In fact, intercropping or rotation of crops is the best means to maintain the soil fertility. During the growth of the cotton plant, equally important operation is to carry weeding to remove unwanted growths which otherwise would consume part of the nourishment of the main plant

Biodiversity helps in promoting productivity of ecological systems. In this, each species plays an important role. If there are more number of plant-species cultivated around the cotton plant, there is greater variety of crops. This increases the sustainability of all such plant species. Ultimately, when biological diversity is improved while growing organic cotton, there is a reduced risk of damage due to insects or even birds & mammals. In organic cotton cultivation even sprays of beneficial predator (insects) are allowed to control the pest naturally. Normally, chemicals are used to defoliate the plant. This helps in easy machine picking operation without much of vegetable impurities. However, the organic cotton is often hand-picked. This, in a way, helps in reducing vegetable trash in bale cotton. As a food material, the seeds of organic cotton are very hygienic. The oil from the seeds (linseed oil) is very rich in proteins. It is also popular oil for frying many eatable items.

2.6.1 Control on Adverse Effects by Growing Organic Cotton[5]

Cotton is cultivated on major area of the total cultivated land in the world and uses not less than 65% of the pesticides. This proportion is alarmingly high as compared to any other single major crop. The pesticides, herbicides, insectides and defoliants are all used at different occasions during the growth of cotton. Some of them depend upon the need and severity of pest attack. The various types and proportions of chemicals used in cotton cultivation from time to time are bound to have their adverse effect on the environment. The sprays which are used on the cotton plantation to control pest are very harmful and are very likely to be released into the surrounding atmosphere. This again pollutes not only air but also surface water. If the residual chemicals continue to remain on cotton fibres, even in minute quantity, through their journey from fibre to finished fabrics, they are very likely to cause discomfort to the customers. Thus, the use of pesticides not only curbs the beneficial biodiversity but also greatly intrudes on existing eco-systems.

By using the healthier farming practices, it is possible for the farmers, especially those growing cotton, to protect the biodiversity. This will enable them to grow and offer healthier cotton product. It will be a great help in making the surrounding atmosphere better. The farmers can contribute by – carrying-out surface protection by planting vegetation, maintaining groundwater quality (no chemicals), using alternative means for insecticide, growing beneficial habitat surrounding the main crop, conserving biodiversity and totally avoiding harmful chemical sprays.

When the crop is grown by adopting organic means, the soil becomes rich. There are more proteins and useful organic matters within the soil and hence it becomes more fertile. By plantation, the soil erosion is also considerably controlled. However, certain changes are required, when growing organic cotton. The land is required to be cleansed at least for 3 years. No non-permitted substances should be used on the land during this period. Only after three years, the seeds of organic cotton can be sown. The plantation must have a proper, strong fencing so as to keep the organic crop plantation from coming in any contact with neighbouring crops. For maintaining and improving the fertility of the soil, use of any kind of chemicals needs to be strictly avoided. The fertility should be improved only by cultivation practices. All possible precautions should be taken to avoid or control soil erosion. For controlling pest, such measures as would help both the crop and the surrounding living organisms, need to be taken. This is done by managing ecosystems. Only crop nourishment (purely organic matters) is allowed during the whole cultivation. The change in the soil structure (some loosening of the surface) is also allowed. This is because with small change in the surface structure, soil is able to hold nutrients and moisture more effectively.

It is a common belief that much less water is required by conventional cotton as compared to organic cotton during their cultivation. In many instances the opposite has been true. When the soil is healthy, no intense irrigation is required for organic cotton. As the plants are grown in a much better and healthier surrounding, the water is used more efficiently by them. It is also observed that during a phase when there is shift from conventional to organic cotton land, the latter may possibly require a little extra water during transition. However, when the organic cotton cultivation has been confirmed over two to three years, the organic cotton plant starts using normal level water and surprisingly in some instances, it consumes even less.

2.6.2 ORGANIC VS. CONVENTIONAL COTTON[1,7]

Until almost 70 years back, cotton growing involved old techniques which neither caused any harm to the natural resources nor led to any hazards to health of human being. Presently in the world, the cotton cultivation is hardly done without using pesticides. This has led to threats from this single factor to the human beings and other wild life thus challenging the safety of the environment. Many of the pesticides used in present cotton cultivation have long been used in the form of nerve gases in the disastrous Second World War. It is observed that when breathed, they cause different forms of cancer. As mentioned earlier, the conventionally grown cotton consumes enormously high proportion of harmful chemical pesticides Some observations were made in U.S.A. and it was reported that the pesticides used for cotton were only a little less than those used for corn – a major crop in US. Some of these chemical-pesticides were far more toxic than well known DDT.

It is interesting to note that a very small percentage of the total chemicals used or sprayed really fulfill the requirements of controlling pest and other unwanted growths. The remaining is all let loose in the air or absorbed by plant or air or water running around the plant. If absorbed by soil or water, ultimately it would reach human beings to harm their bodies. If it is carried by the air surrounding plant, it would pollute the surrounding atmosphere and would cause serious damage to the farmer-workers. This invariably leads to major disturbance in ecological system. Another danger is, in the due course, the insects or pest also develops immunization against such strong chemicals. This forces the farmers to either increase the quantity of chemicals or use still stronger and more harmful chemicals. This is far more disastrous to human beings in the vicinity. It has been reported through one of the studies undertaken in this regard that a huge number of birds get fatal infections every year because of dangerously harmful insectides used to control pest on cotton. In addition, there are other harmful chemicals added owing to the use of toxic herbicides, fungicides and defoliants. Further, during subsequent cotton yarn/fabric processing, chlorine based bleaching agents are used. For 'easy care process' resins treatments are give. All these are equally worse and lead to a variety of health problems. The last to add to these, is genetically modified cotton. Every year, its production has been increasing. Considering its possible use (in the form of seeds) in food items, as explained before, the situation is quite alarming and a positive danger to ecological balance.

Shifting to production of organic cotton has at least some hope of returning to safe and sustainable practices. Grown with natural fertilizers, it is also free from toxic chemicals. Banking on crop rotation ensures maintenance of soil fertility. The use of biological means to control pests and weeds further increase the safety margin. However, the rules for getting the cotton certified as 'organic' are quite strict. No pesticides are to be used in consecutive three years. The subsequent processing of cotton up to yarn or fabric stage has to be according to international standards. There has to be no contamination during transport or storing of organic cotton. In U.S.A., a separate agency is established to issue a certificate for the cotton grown as 'organic cotton'.

Much less carbon di-oxide is released when the farmers adopt manual farming and organic practices. As there are no chemicals or defoliants used in organic cotton cultivation, it is more eco-friendly. It does not have any ill effect on the health of surrounding life. It uses much less water and electricity. There is no added contamination in the form of lead, nickel, pesticides or even heavy metals. This substantially reduces risks to people with allergy or chemical sensitivity when they wear clothings made of organic cotton. Especially, the baby-clothings made from organic cotton are quite safe to wear. As against this, there are serious effects on health with conventionally cultivated cottons. Innumerable cases are reported of serious pesticide-related problems e.g. deficiencies from birth. Some of them are really serious and lead to paralysis or even death. This is because; there are many lacunae in methods of storing, or their handling and transportation.

Just to maintain the yield level, the farmers are always tempted to use more proportion of expensive pesticides. When the expected yield is not received, this puts them in heavy debt, and this leads to a vicious circle. Some times, all this, leads to their committing tragic suicide. Comparatively, organic farming is not only cheaper but healthier also. It promotes socially acceptable alternative to a large-scale farming.

2.6.3 Awareness in Restricting Pesticides

In growing conventional cotton, when the pesticides are used, it leads to serious ailments like giddiness, vomiting, loose motions, irritation, watering of the eyes and blurred vision. It was reported that in Vidarbh (a region in Maharashtra state, India), many farmers died owing to pesticide poisoning. Some of the pesticides like Aldicarb, or Phorate are highly toxic. The other pesticide also used in cotton fields is DDT. Many of such pesticides are persistent. Even when organic cotton field, during the phase of transformation, takes minimum 2–3 years for a complete change-over, the persistency of these dangerous chemicals can still be traced in this cotton. It is observed that to make a simple T-shirt, a large amount of chemicals are needed to grow equivalent cotton.

Through campaigns, many research institutions are trying to orient farmer's community about organic cotton. They help them to slowly switch-over to organic cotton cultivation with a sound biological approach. The Cleaner Cotton projects are being launched and they promises cotton production with far less chemicals. The farmers are also taught the importance of reducing chemicals and following biological farming. Ultimately, the ecological goal is to convert fields from chemical controls to biological controls. Presently, all the organic cotton growers are saving money on production cost and greatly reducing the health hazards. In return, both, the farmer and consumer are not exposed to the chemicals used in conventional cotton farming. Along with cotton, organic models of production and products are made available for purchase at selected locations. The most popular amongst these are baby clothes and diapers which are needed to be absolutely safe to wear.

2.7 COTTON HARVESTING

Most of the cotton varieties in the the United States, Europe, and Australia are harvested mechanically, either by a cotton picker, a machine that removes the cotton from the boll without damaging the cotton plant, or by a cotton stripper, which strips the entire boll off the plant. The cotton strippers are used in regions where it is too windy to grow varieties of cotton, and usually after application of a chemical defoliant or the natural defoliation that occurs after a freeze (cold). The cotton is a perennial (everlasting) crop in the tropics and without defoliation or freezing, the plant continues to grow. The machine picking is much faster and specially employed where the cotton is grown in quite large fields.

The cotton continues to be picked by hand in developing countries such as Uzbekistan and some parts of India. Though hand picking is a very slow process and involves lot of labour complement, in some countries, where extra care is required to be taken to avoid foliage accompanying the picked up cotton or the period of ripening of cotton bolls is spread over a week, the hand picking may be adopted.

2.8   MOLECULAR STRUCTURE OF COTTON FIBRE

Basically cotton contains 91% of cellulose, 7.8% water and the remaining is protoplasm, pectin, waxes, fatty substances and mineral salts.

As mentioned earlier, cotton is basically a chain of glucose units. During linking, two OH groups join and form water. The linking of two residual molecules - called as 'cellobiose' - is as shown in Figure 2.2, in that they turn through 180°. These form the main foundation for cellulose.

The process of condensation (joining of the molecules) continues and a much longer chain is formed - 'condensation polymer'. The Degree of Polymerization (number of repeating units) in cotton is very large (2000–4000), while with regenerated cellulose (Viscose), is only 300–400. Whereas alkalis have no damaging effect, acids can reverse the condensation process, reducing the chain length and damaging the fibre.

2.9   COTTON PLANT & GENUS

As mentioned earlier, cotton is a seed hair. The fibre grows inwardly in the form of a single cell from an epidermis. Cotton belongs to order of 'Malvaceae' or

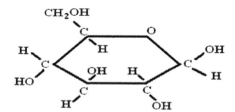

FIGURE 2.2   The Basic Glucose Unit & its link-chain[9,10]: The linking results in forming a long Cotton fibre.

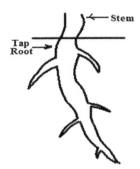
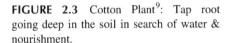

FIGURE 2.3 Cotton Plant[9]: Tap root going deep in the soil in search of water & nourishment.

FIGURE 2.4 Shape of leaf[9]: Typical cotton plant leaf with pointed ends.

'mallows' and its generic name is "Gossipium". Gossipium Herbaceum includes most of the Indian and Russian cottons. Gossipium Arboreum or 'tree cotton' is Asiatic cotton which includes "Sacred Indian cotton". Gossipium Hirsutum is so called from the hairy character of the plant – its stem, leaves and seeds. The American upland (Cambodian) represents this variety. Gossipium Barbadense includes –Sea-Island, Egyptian, Peruvian and other cottons. These are the best varieties available in the world. They are roughly distinguished as 'vine-leaf' cottons.

In appearance and growing characteristics, the plant varies greatly in different countries, but generally it is a bushy plant. The height is about 3–6 ft; with wide spread branches, especially on the lower portion of the stem. At the top, it is slightly tapering. When closely planted, the branching is reduced and it grows more like a raspberry bush.

The root system (Fig. 2.3) varies greatly depending upon the nature of the soil and water supply. In Egypt, for example, the main tap root has been found to descend over 6 ft. into ground in search of water. Normally cotton grows well in deep soils, however, in Barbados; it also grows well on shallow soil giving world's best cotton.

The leaves (Figs. 2.4 & 2.5) are large and more deeply divided into 3–5 lobes, the form again varying with different varieties and species. Sea Island has very deeply cut lobes while American up-land cotton represents the opposite extreme, the leaves being more hairy to give dull appearance. The leaf of Indian cotton is smaller with lobes characteristically rounded in appearance.

The flowers (Fig. 2.6) also differ considerably in colour from one specie to another. In general, the flower is more tubular and surrounded by three large bracts.

The flowers of Sea-Island and Egyptian types are lemon or golden-yellow with crimson spots at the base of the five petals and golden brush of stamen.

Flower

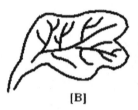

[B]

FIGURE 2.5 Shape of leaf[9]: Cotton plant leaf with rounded lobed leaf.

FIGURE 2.6 Cotton flower[9]: The flowers are pinkish yellow.

American up-land plant, on the other hand, has creamy flowers without any markings and has buff colour stamens. The flower of Indian cotton plant is yellow and smaller than that of Egyptian and spots are larger and darker. The flower of sacred Indian cotton is red.

If the atmosphere is humid, the flowers turn pink on fading and finally almost red before withering out. The flowers usually remain open for 2–3 days. After the fertilization, the flowers turn into a bud-like structure called 'Boll' (Fig. 2.7).

The boll or the fruit, before maturity is of varying shades of green and again varies greatly in size (3/4" to 1½" in dia.) and shape. Thus, Sea-Island and Egyptian bolls, especially the former, are narrower and pointed; whereas the bolls of typical American up-land are much rounder and shorter. The pod is formed with 3–5 compartments within it. Each such compartment bears many seeds. From the epidermis of each seed, the numerous tiny hair like structures start spring-out. These are cotton fibres. In the due course, the fibre mass starts steadily growing. At one stage, the fibre mass can no longer be contained within the pod (or cotton boll), when the bolls burst open (Fig. 2.8).

FIGURE 2.7 Cotton pod[9]: They encase numerous cotton seeds over which cotton fibres grow.

A small cotton field in Cupertino – Photo by Author

FIGURE 2.8   Burst open Cotton Bolls-cotton blooming-out from burst opened bolls.

2.10   CRYSTALLINE-AMORPHOUS REGIONS & THEIR INFLUENCE

The arrangement of the chain molecules in cotton is not uniform. In some of the regions, the packing density of these molecules is very high; whereas, in some other regions the spacing between them is comparatively more open. Usually, in dense region, the arrangement of the molecules is more systematic and orderly. These are called as "Crystalline" regions. The open arrangement of the molecules is called as "Amorphous" region.

The systematic arrangement of the molecules makes the structure more tight and compact; whereas, the open arrangement makes the structure more penetrable. Often these two regions are small and are found to merge into each other. The smallest basic repeat of molecules in crystalline region of cellulose is formed by four glucose residues, two in each of the two parallel planes and is termed as 'unit cell'. The dimension of this unit cell and molecular weight of the residue decide the density of the natural cellulose which is about 1.59 - a value slightly higher than that of cotton fibre owing to presence of amorphous regions.

The natural cotton fibre has approximately two-third crystalline region and one-third amorphous region. As against this, the regenerated cellulose (Viscose) has almost the reverse proportion. It may be mentioned here that it is the crystalline region which gives the strength to the fibre; whereas, amorphous region decides the absorption capacity or dyeability. Thus, the penetration of any liquids (chemicals) is easy in amorphous regions and leads to swelling. In some instances, a small penetration in some of the crystalline regions opens-up the structure (observed by less wet strength).

It has been observed that the chain molecules in different types of cellulosic fibres differ in their alignment with the fibre axis. Bast fibres such as flax, ramie, hemp etc. are highly oriented, cotton being medium oriented while viscose has low to medium orientation; this again depends upon the drawing operation during its wet-spinning.

2.11 BI-REFRINGENCE

All cellulosic fibres exhibit strong "bi-refringence". When a polarized light is passed through the fibre, two different values of refractive indices are obtained - one parallel and the other perpendicular to the fibre axis. The difference between these two values is, in fact, called as birefringence. It is possible to quantitatively assess the degree of orientation. The ratio of value of birefringence of a fibre to that estimated for a fibre with perfect alignment to the fibre axis helps in this case. If this ratio is closer to zero, it indicates "random" orientation; whereas, the one closer to unity indicates high orientation. The study also reveals that the crystallites are inclined to the fibre axis. With best varieties of Sea Island cotton, this angle is comparatively smaller (about 25°) while with coarser Indian varieties it is higher (about 35°). It is also found that there is better orientation with longer and finer fibres; whereas it decreases as both the fibre length and fineness decrease.

The fibre is fairly uniform in width in the middle section while around end portion, it varies in its diameter which ranges from 12–20 micrometers. The length depends upon the variety and usually ranges from 22 to 36 mm. The fibre has moderate to high lustre. The tenacity varies depending upon whether tested in dry or wet conditions (Dry – 3.0-5.0 g/d, wet - 3.6–6.0 g/d). The tenacity values with Stelometer, in g/tex at 1/8 inch guage indicated their strong correlation with the fibre length [weak fibre (20 mm and below); average (20-29 mm) and very strong fibre (above 32 mm). The density of the fibre varies between 1.54 to 1.56 g/cm^3, whereas; its resiliency is on the lower side. The cotton fibre has good absorption capacity; its natural moisture regain is around 4–6% whereas under saturation condition it can withhold 8–10% moisture. When mercerized, the moisture-holding capacity further increases. Dimensionally, it is a reasonably stable fibre.

It possesses fairly low elasticity, exhibits bi-refringent property (0.046) and the normal micronaire range is from 2 to 6.5. Fibre fineness values in millitex range from 135 or lower (very fine) to 230 and above (very coarse). Whereas, the corresponding denier-values are 0.7 and 2.3, the frictional coefficient is 2.5 the value of which changes rapidly when treated or when wet. When maturity ratio is below 0.7– 0.8, it is considered to be immature; whereas anything above 0.9 is very mature fibre.

While it is fairly resistant to the alkalis, the acids damage and weaken the fibre. It is again fairly resistant to all organic solvents. Under the prolonged exposure to sunlight, the fibre weakens. It is susceptible to attacks from mildews, bacteria and insects. At 150 °C it decomposes with heat and with flames burns readily.

2.12 AGRICULTURAL PRACTICES

In case of cotton plant, several factors like quality of the seeds, their multiplications, new strains resistant to diseases, cultural practices followed, protective measures etc. - all affect and influence the growth. The experiments carried out in agricultural research focus their attention to increasing yield, quality and in building resistance to attack by pest, bacteria and other diseases. It is possible to improve quality by cross-breeding (hi-breed varieties), whereas in some other cases, varieties resistant to droughts are developed. Some of the factors affecting the overall production and quality of the cotton grown are mentioned below.

2.12.1 SEEDS

The seed being the basic material for cultivation, quality of the seed is very important. The research organizations work for both developing the new varieties and making them available to the farmers in sufficient quantity. The seed multiplication in the first stage produces seeds of good quality. In the successive stages of multiplication, however, the quality deteriorates. The size of the seed is equally important - bigger and heavier seeds give higher yield.

2.12.2 AVAILABILITY OF WATER

Whether the crop depends upon rain water or is supplied with irrigation water are decisive factors. This is because; the cotton crop is sensitive to moisture in the soil. Rather than the total quantity, timely supply of water at different stages is very important as it decides both quality and quantity. Though more frequent watering results in bigger bolls with more seeds, excessive or inadequate watering (moisture content in soil) can adversely affect fibre properties. Untimely rain, especially at the time of ripening or picking of bolls can be harmful; whereas, during maturation period, light drizzles with warmer temperatures are found to be beneficial for proper thickening of fibre wall.

The water scarcity during sowing and during the cotton growing season leads to fibres with lower maturity and hence weight. It is known that frequent and timely irrigation facilities do not significantly affect fibre length; but the fibre weight and maturity are better and the neppiness tendency is reduced. Owing to higher fibre weight, however, the yarn strength is slightly reduced. Nitrogen dose along with adequate supply of water is found to be beneficial for cotton quality.

2.12.3 MANURIAL TREATMENT

Along with suitable climate, fertile soil helps the growth of the plant. While climatic conditions are beyond control, the condition of the soil can be improved by judicious application of manures. The object of using the fertilizers is both to make-up for the deficiency as well as to neutralize those constituents which are in excess. Therefore, the application of the manures to already fertile land does not increase the yield. When the deficiencies are made-up, however, the production increases significantly.

Adding fertilizers to a fertile soil marginally reduces the mean length. With low fertile land, addition of nutrients in the form of fertilizers results in stronger and yet neppy yarns. Using nitrogen salts (ammonium phosphate or super phosphate) also has to be done with care; otherwise mean length, fibre weight and fineness are affected. With the development of cotton bolls, the nitrogen content in the stem and leaves reduces. The nitrogen dosage at this time can improve the development of the fibre.

2.12.4 CULTURAL TREATMENT

Loosening of soil brought about by ploughing, preparation for cultivation, seed sowing rate, their spacing, date of first sowing, mixed cropping etc. can all influence the yield as well as fibre properties. Ploughing does not seem to have effect on ginning percentage, fibre length-fineness and fibre maturity. The seed spacing also has marginal influence on length and fineness. With higher seed rate, the fibre seemed to be finer. Early sowing resulted in longer fibre; whereas the adverse effect of late sowing seemed to be partially compensated by heavy irrigation. Mixed cropping (two crops taken at a time) sometimes is found to give better ginning percentage along with longer fibre length.

2.12.5 FLOWERING & FRUITING

The duration of flowering and fruiting also varies considerably and this has influence on the characteristics of fibre. This is because, when the picking of cotton gets extended, the required exposure to sunlight also varies. Unless manual picking is employed, there is always a risk of under or over exposure. As against this, with less exposure, there is a risk of cotton being picked-up with sufficient moisture left over and it becomes more susceptible to microbial damage during its subsequent storage. Over exposure also leaves the cotton open to hazards owing to vagaries of nature and sudden changes in the atmospheric conditions.

Flowering period also varies from plant to plant and hence all the bolls do not mature at the same time. Usually, the bolls formed due to late flowering do not get sufficient nourishment from the soil and thus lead to poor quality of the fibre.

2.12.6 GENERAL CONDITIONS

Some varieties of cotton when grown in certain locations show different yield and quality. This is basically influenced by the soil characteristics and the climatic conditions. While choosing the variety, therefore, the one which shows minimum difference in yield and quality, when sown at different places, is considered to be more suitable owing to its adaptability.

There are typical synthetic hormones which have beneficial effects in the development of cotton and influence yield. These are called as growth promoters and Regulators. These are scientifically formulated with most vital nutrients that increase the crop growth through plant metabolic activities. They are in the form of sprays comprising of amino acids and other nutrients that enhance crop output sometimes by as much as 30%.

These hormones increases the uptake of micronutrients, stimulates the properties, promotes enzymatic activities and accelerates photosynthesis. They promote protein and amino acid generation to act as a building block for proper growth. The hormones activate and induce flowering, and improve the efficiency of pesticides and fungicides by increasing internal immunity. They also increase the crop health-resistance to draught. They further helps in strengthening the roots and their spreading, thus helping the plant growth.

2.13   COTTON FIBRE AND ITS MATURITY

Structure-wise, cotton fibre has a broad base at its root with tapering apical end. During the growth, it is almost round in shape. Later, during the exposure to sunlight after the bolls burst open, the inside mass dries and the fibres collapses into a flat ribbon with convolutions. These convolutions are all along its length and often reverse their direction.

When the fibre is looked at low magnification (1500x), there are only two distinct visible regions – the outer wall and the central lumen (hollow portion). However, at much higher magnification (20,000x) with action of swelling agents, the wall is further differentiated into: (1) Primary wall and (2) Secondary wall (Fig. 2.9). The former shows no details under ordinary light, however, it appears to be a mesh of fine strands when stained with Congo-red and observed under polarized light. These strands are spiraled to left and right at 70° to the fibre axis. There are transverse strand as well, when viewed at an angle of 45°between the planes of cross Nichols and fibre axis. It is only the primary wall which comes into contact with the machine components when the fibre is processed. It also plays a vital role during drafting and twisting; whereas the latter (secondary wall) gives the strength to the yarn.

The primary wall has an outer coating -'cuticle' - and it contains the mixture of fats, waxes and resins released at the surface of the cell during maturation.

The primary wall is composed of protein wax and cellulose. The high magnification under polarized light also reveals the structure of secondary wall. It consists of layers of cellulose. These layers vary from fibre to fibre, depending upon the duration of maturation period.

Each layer represents a growth ring called 'Lamella' and is composed of many disconnected fibrils which again follow a helical path, with several changes in the spiral angle to the fibre axis. These fibrils are laid in the radial alignment with those in earlier rings. The secondary wall is therefore, perforated with cracks or pits. The internal structure of these fibrils is essentially native cellulose.

The thickness of primary and secondary wall therefore, decides the material (cellulose) content in a fibre. If the walls are completely and fully grown, the fibre is fully matured. In such cases, the size of the central canal - lumen is very much smaller. Depending upon the size of the lumen and the wall thickness, the fibres are classified with the following procedure.

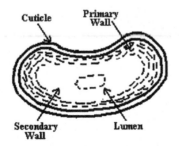

FIGURE 2.9   Cotton fibre under microscope[9,10]: At high magnification, distinct regions are visible.

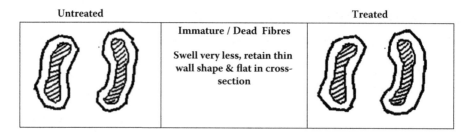

FIGURE 2.10 Fibre maturity[9]: Treatment with 18% NaOH makes the fibre swell. The fibre cross-section before and after the swelling is shown.

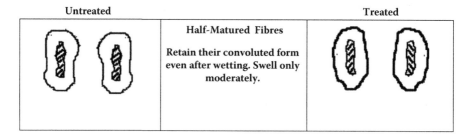

FIGURE 2.11 Fibre maturity[9]: Treatment with 18% NaOH makes the fibre swell. The fibre cross-section before and after the swelling is shown.

The fibres after soaking in 18% NaOH are observed under the microscope with approximately 45–50x magnification. If there is a very small deposit of primary & secondary wall with a much bigger size of lumen, the fibre s termed as 'Dead' or Immature fibre (Fig. 2.10). When there are varying deposits of secondary wall and primary wall and a comparatively larger space is occupied by lumen, the fibres are called as 'Half mature' (Fig. 2.11). With full deposits of secondary wall and primary wall, leaving hardly a space for very small insignificant size of lumen, the fibres are termed as 'Fully mature' (Fig. 2.12).

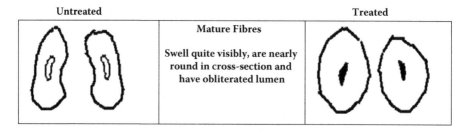

FIGURE 2.12 Fibre maturity[9]: Treatment with 18% NaOH makes the fibre swell. The fibre cross-section before and after the swelling is shown.

If the diameters of the two fibres are the same, their weights would vary depending upon the level of their maturity, the higher weight showing higher maturity. Obviously, due to their cellulosic content, the matured fibres are stronger and contribute more to the yarn strength; vice-versa, immature or dead fibres are very weak and hence do not contribute their share of strength. In fact, they are more prone to breakages during the mechanical processing and also appear as the short fibres in the subsequent stages. These short fibres are source of trouble in drafting process and they try to protrude out of the body of the yarn making it appear more hairy.

Another way of classifying these fibres is by viewing them under polarized light under cross Nichols (angle of 45° to polarized light). Mature fibres appear to be yellowish, semi or half mature appear to be green, whereas immature or dead fibres have bluish or purple tinge shade.

2.14   FIBRE PROPERTIES & SPINNABILITY

The cotton breeders are always after developing cotton varieties as will be useful to the textile industry. This is because, the fibre properties directly reflect into the yarns and later into fabrics. In general, the potential of any cotton is judged in terms of the yarn which is made from it, especially its uniformity and strength. Alternately, depending upon this potential, the yarn can be profitably spun to a certain count with the acceptable strength and the uniformity. This count is referred to as 'Spinning Value' of the cotton. There are other factors too, like appearance, neppiness, imperfections, ease of processing and end breaks in the final spinning of yarn.

When the yarn breaks due to stress, only some fibres break; whereas, many others merely slip, thus involving two important fibre properties – the strength of the fibres and their clinging power. The fibre length, however, seems to be most important factor contributing to the 60% of the yarn strength whereas, fibre strength accounting for hardly 15–20%. Many research workers have concluded that fibre strength and clinging power do not show marked association with spinning value. However, 'Balls' (Studies of cotton quality) stated that from the point of view of yarn strength, the most important fibre properties are - intrinsic strength, fineness and slipperiness (opposite of clinging power).

The fineness of the fibre is equally important. Its influence, together with fibre length on yarn strength is very high. The spinning value of the cotton is, therefore judged from these two factors. In majority of the cases, the longer cottons are also finer. The maturity is yet another important factor. Increase in maturity, especially for shorter and coarser cottons makes them more rigid, making the fibre weight higher. However, for longer cottons, improvement in the maturity improves their spinning value.

2.15   CONCEPT OF HIGHEST STANDARD WARP COUNT (H.S.W.C.)

It is therefore, difficult to relate in a simple equation form, the contribution of different fibre properties towards the final yarn strength. Even then, a term *Highest*

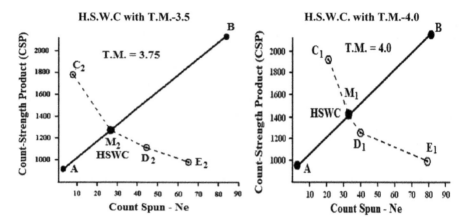

FIGURE 2.13 Highest Standard Warp Count (H.S.W.C)[7.9]: It is the count which can be most beneficially spun from a given cotton . It indicates how fine acotton can be spun with useful yarn properties.

Standard Warp Count had been introduced by which, it was possible to relate the fibre properties with the yarn strength. It is known that the finer the cotton and the longer it is in length, it can be spun to finer counts. On this basis, standard values of the potential strength (C.S.P. = count × lea strength) for other various counts are plotted (line AB in the graph Fig. 2.13). Though the standards are arbitrary, they represent the average quality of the types of cottons used in the mills.

The line AB indicates the expected CSP for different count. It can be seen that as the count becomes finer, the CSP values increase, thus indicating higher quality requirements from finer counts.

Unknown cotton, to be tested for its spinning value, is spun to a certain count in vicinity of its potential, the potential being estimated depending upon fibre characteristics, especially, length and fineness. The twist multiplier (say 4.0) used to spin the yarn is again related to the count range.

Two more counts are spun with the same twist multiplier around the count chosen earlier. The CSP values ($C_1D_1E_1$) are thus found for the three counts so spun. The point of intersection of AB and $C_1D_1E_1$ is the point M_1, which is the highest standard warp count.

It is evident from the graph that if any cotton is spun to a count lower than HSWC, the CSP of the yarn would be greater than the standard for that count. However, if it is desired to spin higher than HSWC, there will be proportionate reduction in strength, i.e., some strength will have to be sacrificed.

Equally interesting is the fact that when these three counts are spun with a lower twist multiplier (say 3.75 – Fig. 2.13), the graph changes to a new position - $C_2D_2E_2$ and the new point of intersection is M_2, a point representing lower HSWC. Thus, the value of HSWC also depends upon the twist multiplier used. Therefore, while comparing the two cottons, belonging to the same class, it is necessary to use same twist multiplier to compare their HSWC values.

The highest standard warp count is not the highest count to which the yarn can be spun. In fact, the graph clearly reveals that, when the yarn is spun to counts higher than HSWC it still exhibits certain strength. However, this strength would be lower than the expected standard value. The yarn thus would be weaker and possibly may not perform well when spun or later woven into fabric. Even then, as a single index, HSWC is definitely useful to the cotton breeder as a guiding tool. It may again be noted that, since this value is related to the twist multiplier used, higher strength of the yarn may be achieved while spinning to count higher than HSWC, but only at a higher cost (production rates in ring frame are lower). Alternately, it may be possible to spin the higher counts without increasing twist multiplier. Such yarns will be softer, weaker and possibly could be used as weft yarn. ICAR (Indian Council for Agricultural Research) has published data for fibre properties of some typical cotton varieties grown in India.

TABLE 2.4
Fibre Characteristics & Spinning Value of Typical Cottons

Cotton	Span Length 2.5% (mm)	Micronaire Mc	Bundle Strength at 3 mm – g/tex	Trash %	Spinning Value
MCU-6 Bharati	25–26	4.2	23–25	3.5–4.5	30s–34s
F-414	26–27	4.3–4.5	24–25	4.0-5.0	28s–34s
LRA	28–29	3.6–3.7	24–25	2.5–3.5	30s–40s
Mech-1	29–30	3.4–3.7	25–27	3.5–4.5	34s–44s
MCU-5	32–34	3.2–3.3	27–28	2.5–3.5	50s–60s
Shankar-4 (H-4)	28–29	3.3–3.6	23–25	3.5–4.5	38s–44s
Shankar-6	28–30	3.6–3.8	23–24	3.1–3.4	38s–44s
G-27	16–17	6.6–6.8	47–48	5.0–6.0	Below 10s
Jayadhar	22–23	4.4–4.8	23–24	4.5–5.0	18s–20s
Sanjay	23–24	4.5–4.6	25–26	5.0–6.5	20s–24s
A-235	23–24	4.8–5.0	25–27	4.0–5.0	20s–22s
Y-1	25–26	4.5–5.0	23–24	4.0–5.0	20s–24s
DCH (Jayalaxmi)	34–35	2.8–3.0	28–30	3.0–4.5	70s–80s
Suvin	36–38	3.0–3.2	36–37	2.0–2.8	100s–120s
Jyoti	24–25	4.5–4.8	22–23	3.5–4.0	20s–22s
Laxmi	24–25	3.8–4.2	22–23	3.5–4.5	20s–24s
Khandwa-2	24–25	4.0–4.1	23–24	3.5–4.5	28s–30s
J-34	25–26	4.0–4.5	23–25	4.5–5.5	22s–26s
MCU-7	25–26	3.6–4.0	23–24	3.0–4.0	24s–28s
Krishna	25–26	3.9–4.0	24–25	3.0–3.5	34s–36s
Buri	26–27	3.6–3.8	45–46	–	30s–40s
Varalaxmi	29–31	3.0–3.1	44–45	–	60s–80s

TABLE 2.5
Some Standard Varieties of Cotton in India

Standard	Staple mm	Micronaire	Strength g/tex	Trade Name	Where Grown
ICS-101	<22	5.0–7.0	15	Bengal Desi RG	Punjab, Haryana, UP, Assam
ICS-102	22	4.5–5.9	19	V-797	Gujrat
ICS-103	23	4.0–5.5	19	Jayadhar	AP, Karnataka
ICS-104	24	4.0–5.5	20	Y-1	MP, AP Maharashtra
ICS-202	25	3.5–4.9	23	J-34	Punjab, Haryana, Rajasthan
ICS-105	27	3.5–4.9	24	LRA-5166	MP, Maharashtra, AP, Karnataka, TN
ICS-105	28	3.5–4.9	25	H-4/MECH 1	Maharashtra, MP, Karnataka, AP
ICS-105	29	3.5–4.9	26	Shankar-6	Gujarat
ICS-105	31	3.5–4.9	27	Bunny-Brahma	Maharashtra, MP, Karnataka, Orissa, TN, AP
ICS-106	33	3.3–4.5	28	MCU-5 Surabhi	Karnataka, TN, Orissa, AP
ICS-107	35	2.8–3.6	31	DCH-32	Karnataka, MP TN, Orissa, MP

2.16 FIBRE QUALITY INDEX (FQI)

As mentioned earlier, it is known that few important fibre properties decide the potential of cotton. The commonly used fibre properties are - (1) Effective Length (or 2.5% Span Length) (2) Uniformity Ratio (UR) (3) Bundle Strength and (4) Micronaire. There are some other properties like gravimetric fineness, trash content, colour, maturity, and tenderness which directly or indirectly do control the quality of cotton. Even then, the above four properties more or less completely convey the cotton potential for spinning a yarn to a certain count.

The concept of Fibre Quality Index (FQI) combines all these four properties so as to represent a common index. FQI is thus a function of these properties. The logic in establishing this relationship lies in the influence that these properties have on yarn quality. For Example, it is known that both effective length and bundle strength would have positive relation with yarn quality i.e. higher the values of these, the better will be the yarn formed. UR is the ratio of 50% span length to 2.5% span length. Therefore higher is this ratio, it positively influences the yarn quality. The Micronaire (Mc), on the other hand, represents combined influence of both fineness and maturity. Hence, when one variety of cotton coming from different places (stations) is tested, the lower micronaire value would indicate lower maturity of that lot; whereas, when two altogether different cottons are tested, a lower micronaire would indicate finer cotton. In general, lower the micronaire, the finer is the cotton. The FQI is defined in different ways.

$$FQI = \frac{2.5\% \text{ Span Length} \times \text{U.R.} \times \text{Maturity Coefficient} \times \text{B.S.}(1/8'')}{Mc} \quad (1)$$

It may be noted here that an additional variable - Maturity Coefficient is used in the above equation. The value of Bundle Strength is as obtained from Stelometer at 1/8 inch gauge. In fact, there are two values for Bundle Strength - one at '0' gauge and another at 1/8 inch. However, it is observed that the latter is better correlated with cotton potential. Work carried out at ATIRA reveals that a product of the two strength values from Stelometer are still better correlated with lea strength. With this FQI can be expressed as follows:

$$FQI = \frac{2.5\% \text{ Span Length} \times \text{U.R.} \times \sqrt{[(\text{B.S. at } 1/8'' \times \text{B.S. at 0 gauge})]}}{Mc} \quad (2)$$

$$FQI = \frac{\text{E.L.} \times \sqrt{((\text{B.S. } 1/8'' \times \text{B.S. } 0''))}}{\sqrt{(\text{fibre weight/length} \times \text{Std. fibre weight}}} \quad (3)$$

Where,

B.S. = Bundle Strength from Stelometer

The expression (3) is based on the fact that Micronaire is a function of the product of maturity ratio and weight per unit length. The concept of Fibre Quality Index is based on a sound logic. This is because there exists an inter-relationship between various fibre properties. Therefore, when taken together, they do represent the cotton potential in the form of yarn parameters, especially the yarn strength.

However, it is quite possible that the two cotton differing in their properties can still give the same fibre quality index, the deficiency in one property being compensated and made-up by other. Therefore while preparing a mixing, it may be possible to choose a certain variety of cotton slightly inferior in length and compensate it by choosing another variety which may show superior bundle strength. It may also be remembered that the relative effect of individual property is not going to be the same on the yarn quality. Therefore, this aspect will have to given due weightage while deciding the proportion of mixing.

The effect of these various properties on the performance of the final yarn spinning and post spinning processes is another important criterion. It is thus, possible that while one of the constituent in the mixing is poor in maturity, the mixing as a whole, may still be equal in its potential as regards the strength of yarn. However, the yarn may be inferior in some other respects e.g. the imperfections, especially neppiness.

In this respect, therefore, the fibre quality index has its own limitations in completely describing the cotton potential. Even then, as regards the yarn strength, the quality index gives quite useful information and guidelines. Bogdan (T.R.J., 26, 1956, p. 20) has come out with a parameter 'p' which has been related to count, lea strength, twist multiplier and F.Q.I. A close relationship was found to exist between 'p' and these parameters.

2.17 AGRICULTURAL RESEARCH IN INDIA

India is the only country in the world that grows on a commercial scale all four cultivated species of cotton viz., Gossypium Arboreum, Gossypium Herbaceum, Gossypium Hirsutum and Gossypium Barbadense. In the country therefore, the cotton is a premier cash or commercial crop. It has enormous potential in generating employment in rural and urban sector and at the same time, developing economic and trade activities both within and outside the country. About 60 million people depend on cotton cultivation, trade and processing. Indian cotton meets diverse requirements of mills, power-looms, handlooms and cottage industries.

Significant improvements have been made in cotton-lint yield after independence (from 81 kg/hectare to 374 kg/hectare). Along with this, the average cotton seed yield also has increased. Apart from length, strength, fineness, maturity and uniformity of the fibre, the spinning value of the cotton is also important. MCU-5 was the first extra-long staple variety of Hirsutum type released in India. Sujata was another extra-long staple variety in Barbadense type and it made a significant breakthrough in superfine spinning (finer than 100s) and was comparable to Egyptian cotton. The release of another variety 'Suvin' is yet another land mark in superfine cotton varieties having a potential up to 120s count. It is grown in Tamil Nadu and is comparable to Giza-45. More long staple varieties – K-8, K-9, K-10, K-11, PA-183, Arvinda and AKA-8401 have been also released.

The cotton crop is very often attacked by several diseases and insect pests. This results in considerable loss in both quality and yield. Only genetic resistance is the solution in reducing such losses. A variety –MCU-5 is developed to resistant to Verticillium wilt. There are other varieties - B-1107, SRT-1, Khandwa-2, DHY-286 & PKV-081- all of which are resistant to fusarium wilt and jassid. They have been genetically modified and released. On the same lines, LK-861, Supriya and Kanchana are developed to resist white fly.

Some varieties were developed to have wider adaptability e.g. Bikaneri Narma (in Punjab), MCU-5, LRA-5166 (both in Tamil Nadu) & SRT-1 (Gujrat). Some of these were originally developed in certain state, but later, were also available in the other states.

There are often certain production constrains affecting productivity of cotton. They are - multiplicity of varieties & hybrids, cultivation of non-descript genotypes or cultivars (plant selected with desirable characters), low application of inputs (fertilizers, insecticides), imbalanced use of fertilizers, improper spacing in the plants, attack of boll-worms & sucking pest, inadequate rain water & irrigation management, sowing delayed due to start of the monsoon and salinity-excessive moisture-shallow depth of soil.

India has been the leader for developing hybrid cottons (Nearly 200 varieties & hybrids released). The breeding has been mainly focused to increase resistance or tolerance to many pests and diseases. There are short duration varieties like PKV-081, Anjali and hybrids such as Fateh, LHH-144, Omshankar, Dhanlaxmi and Maruvikas. All of these help in multiple cropping according to zonal suitability. It should be mentioned with pride that some superior cotton varieties like MCU-5,

Sharada, Suvin, Varalaxmi and DCH have been developed since long time and have fully settled in the country, thus leading to self-sufficiency.

Hybrid-4 can be spun to 50^s; Varalaxmi can be spun to 80s. Varalaxmi was later replaced by DCH-32 which gave better yield and early maturity. Two more varieties – TCHB-213 and HB-24 have been released and they are still better than DCH-32. The other improvements over H-4 are H-6, H-8, AHH-468 & JKHY-1 and they give higher yield. However, only H-6 is better in properties.

Hybrid seeds of cotton have been derived from BT cotton varieties and they have both advantages and disadvantages. Hybrids are generally expensive, though high yielding. They force the farmers to go back to the seed companies for fresh supply every year. These varieties are sturdier, but often are low yielding. There are more than 30 companies in the country that sell BT cotton, the price ranging from Rs. 450–800 per 450–500 g. There are certain cotton growing areas where the hybrids do not grow; but only BT varieties work there. The average yield of cotton has almost been doubled in past few years and this seems to be owing to transgenic or BT cotton.

In India, the area under the cotton is now almost stabilized. The area under the irrigation is also not likely to increase. Therefore, improving the productivity in both rain-fed and irrigated areas is the only way to increase the out-put. Both Hybrids and BT have a very important role to play in this. The Govt. has instituted cotton research to develop more varieties of hybrids at Nagpur, under Indian Council for Agricultural Research (ICAR). Some of the new varieties like PCHH-31 (Fateh), HHH-81 and Raj-16, are recently released and are expected to improve productivity in irrigated areas.

2.18 PACE & POLICY OF COTTON

Every year, Cotton Advisory Board (CAB) gives its estimate of the production of cotton in a year. The actual production however, varies depending upon various conditions. The prevailing temperature is one such factor. The presence or absence of winter-rains and dew are other factors. Especially, an absence of winter-rains and dew leads to acute moisture stress which can seriously affect the yield. At certain places, the pest attack is severe and this directly affects the yield. Maharashtra is one such state where the yield per hectare is significantly low. However, the area under cotton cultivation in Maharashtra is largest. There are two states in India, Gujrat & Andhra Pradesh - where the yield is often reduced owing to damage by floods.

India is one of the major countries, exporting cotton. The government is unlikely to curb cotton exports. However, there is always an export ceiling for every year. The liberalization of this limit depends on the domestic prices of cotton. There is always a pressure from domestic textile industries to curb the cotton export trade. This is because, only then, the domestic cotton supply would be available at an affordable price. The increase in demands from the domestic consumption always helps in reducing the pile of stock. In addition to this, when the cotton prices are stable over a certain period, increased mill usage always improves the yarn price. As against this, the demands for Indian cotton from countries abroad ultimately boost the returns for the farmers. Similarly, when the cotton arrives in domestic

TABLE 2.6

Comparison of Typical Cotton Production in Different States[7]

State	Approx. Area %	Estimated Production %
Punjab	3.50	2.56 3.53
Telangana	16.05	11.22 12.32
Haryana	6.30	4.98 6.57
Rajasthan	4.90	3.74 4.32
Gujrat	24.65	20.38 22.35
Maharashtra	36.05	36.27 38.24
Madhya Pradesh	5.10	6.8 8.35
Karnataka	3.45	3.60 4.32
Total	111.55 lakh hectares	337.25 lakh bales (170 kg each)

market, it triggers the pace for big purchases by textile mills and helps in offering good price to the cotton producers.

On one hand, the exporters of cotton are interested in sticking to schedule of shipping to the outside countries. This is with a view to earn big profit and also the concessions from Govt. against the foreign exchange. However, their dreams heavily depend upon the quality of present cotton. On the other hand, there are demands from domestic market - the cotton mills. They may be interested in seeing that there immediate demands are first met with. Once this is done, they may wait for the subsequent lean period in cotton prices for increasing their profit margin. In short, the cotton prices may fluctuate depending upon the control of the forces mentioned above.

Whether the cotton is exported or used domestically, it is the farmer who is always the sufferer. This is because, once the cotton is purchased by the agents/ co-operative societies/cotton federations or the mill owners, whatever efforts that are subsequently made to earn profits on this cotton never reach back to the farmers. Many a time, he is in a hurry to sell-off his product to earn immediate price for his goods. The most important controlling factor to force him to do this transaction is the debt that he has taken from either the money lenders or the banks. Even after this hurried transactions, he is still not able to get that price which would not only free him from his debt, but also would provide enough money for his family.

In addition to this, many farmers in India lack the knowledge of the scientific and technological information. This is likely to prevent modernization in agricultural practices. However, now large number of agricultural Institutions is steadily coming-up in India. The agricultural Universities are also helping to bring out many budding scientists. They all are expected to reach the technology to the needy farmers. The stagnation in the growth of cotton prevailing in the past is slowly being controlled and overcome. The technology transfer is being slowly but steadily improving. Earlier, in spite of many institutions involved in teaching, research and developments, they had their own system of educating farmers. The reach of the agricultural science to the then benefactors (farmers) had been limited. There were testing facilities for soil, seeds, irrigation water, agricultural chemicals and fertilizers, but they were not situated at one place. Thus, the farmers were required to run from pillar to post simply to discuss the issues at hand or to seek solutions to their problems.

2.19   NEED FOR CENTRALIZING THE FACILITIES

It is now fully realized that these facilities need to be located at one place in towns so that a large number of needy farmers can avail them. It is also thought to establish such centers where 'mandi' yards or vegetable markets are located. This is because; during paddy season quite a large number of farmers visit such places. Therefore, it would be of great advantage to establish "Farmer's Education Centers" under one roof at these places. In these centers, testing of soil, water, seeds, chemicals and fertilizer analyzing laboratories are expected to be housed in one sub-center, whereas identifying diseases, pests and any abnormalities in plant growth would be housed in another sub center under the same roof. There would also be experts to advise the farmers to grow certain crops in their soil so that it becomes profitable. These experts can also expose the farmers to new crop varieties which have greater resistance to diseases and pest attacks. They would teach new techniques of using natural fertilizers. They would educate the farmers on using minimum spray of pesticides and crop rotation technology. All this would immensely help the farmers to upgrade their knowledge and boost confidence. The exhibits of good farming practices for rabbi, kharif, spring-summer crops and plant protection measures can also be put on display boards at such centers.

The farmers can be advised to use appropriate dosage of fertilizers and chemicals to get higher yield with much less use of pesticides. If government takes initiative, in controlling these activities, the farmers will feel that they are not cheated. Young generation of farmers can be trained so that, in turn, they pass on this knowledge to their senior colleagues through the use of latest communication technology.

LITERATURE REFERRED

1. Technical Bulletins ICAR].
2. Wikipedia.
3. Economical & Political Weekly - Suman Sahai ^ Shakkelur Rehman, 20th July, 2003.
4. BT Cotton – 'A Painful Episode', Courtesy of 'Kapas Utpadak Hitrakshak Sangh', Vadodara, Gujrat).
5. Organic Cotton, A nature's gift to mankind – Dr. Shefali Massey & Dr. Shahnaz Jahan– Textile Magazine, Sept. 2010.
6. Information published by Cotton Association of India.
7. Central Institute for Research on Cotton Technology (CIRCOT) –ICAR Publications.
8. Coloured Cotton – Dr. P. Singh, Dr. V. V. Sigh & Dr. V. N. Waghmare – C.I.C.R. (Nagpur) Technical Bulletin No. 4.
9. Elements of Spinning – A. R. Khare, Sai Publication.

3 Picking, Baling & Ginning

3.1 PICKING

As explained earlier, before picking starts, it is essential to ascertain that the cotton mass inside the bolls is fully dried. This is because, apart from extra moisture within the boll, there are other watery food juices during the continuation of growth. The drying is done by exposing the mass to sunlight. Even this needs to be carefully controlled as the cotton is likely to get over dried and as such, the fibres lose their strength. The exposure to the sunlight also helps in allowing some time for the nourishment to reach some under-developed fibres. This, in turn, improves their maturity. On the other hand, the exposure also allows the fine dust, sand and broken particles of leaves to settle on the partially wet mass of cotton. This is basically owing to strong winds blowing across the field.

The cotton can be either picked up by hand or by machines. In both the cases, a certain level of skill is involved. When it is hand-picked, the labourers go around the field picking up the loose hanging mass of cotton from the bolls. During this time, they also ascertain if the inside cotton mass is sufficiently dried and leave those bolls which require some more exposure on this account. The delay is also influenced by all the bolls not bursting open simultaneously. Thus, there could be more than one picking session around. Especially with very long staple cotton, this method of picking is more beneficial. This is because careful picking allows less trash to accompany the picked-up mass.

A crop is considered good for picking when the hanging mass of cotton from larger cotton bolls is loose and fluffy. It can, therefore, be easily picked in this condition. When the bolls are small and tight, picking does not become easy and the seed cotton is required to be rather pulled-off than picked.

When the fields are huge and widely spread, handpicking not only involves more labourers, but also the operation itself is more time consuming. In many African countries, seed cotton is picked by hand, whereas in the the United States, Australia, Europe, Brazil and Uzbekistan, it is picked up by using huge machines – "Cotton Pickers" or "Cotton Strippers". The experience in setting such machines also counts. The modern harvesting machines not only finish the job quickly but also improve the quality of picking. This is further assisted by a special pre-picking process called 'De-foliation' where the leaves on the plant are destroyed.

However, when the fields are not very large and the bolls do not mature at the same time, the use of harvesting machine becomes inconvenient and uneconomical, especially if the labour for picking cotton is available cheaply. Even then, when there is a shortage of the labour, machine picking has to be adopted. On an average, the trash content in machine-picked cotton is always higher than that from hand-picked one. Thus, a varying proportion of twigs, hulls, dried crushed leaves and other impurities gets associated with machine-picked cotton. Hand-picked cotton, being cleaner, requires less cleaning operation in the subsequent stages, and this protects the intrinsic properties of the fibres. On an average, a tonne of cotton (1000 kg) yields 300–350 kg of fibre, 600–650 kg of seed and approximately 40–50 kg trash.

In industrialized cotton-producing countries, after about 80% of the bolls have opened, it is customary to spray the fields with "Ripener" which speeds up ripening. Subsequently, defoliant is also used to make the leaves fall off. This facilitates easy machine picking from the bolls. When almost 95% bolls have opened, the crop becomes ready for picking. This is when the cotton pickers come into picture. The machines can pick up 800 kg/hour of seed cotton as compared to 80–90 kg per day with hand picking. The cotton thus picked up surrounds the seeds and hence is referred to as "seed cotton".

It is surprising to note that even when the normal picking is over, it is really not so. There are occasional instances where the bolls are not ready to be picked up. This is because they opened much later. They are usually hand-picked as they are very few and scattered all around. This is called "Stripping".

In hand picking, the labourers are generally paid according to the quantity of cotton picked. They are, thus, likely to become careless while picking. This allows some vegetable matter getting associated with hand-picked cotton. In India, many hand-picked varieties, therefore, have a little higher proportion of trash. In Egypt and Sudan, where the cotton quality itself is much better, the experts are employed to pick and choose only the matured cotton from the fields. In such cases, there are more than three distinct pickings.

The first picking always gives the best quality of cotton, the subsequent pickings resulting in a slight drop. When there is a considerable drop in quality from later pickings, it is advisable not to delay the process. The vagaries of the nature are also required to be taken into account. In West Indies, there are always untimely rains. If this happens during picking, the cotton becomes dull, discoloured and develops yellow-brown stains. Further, the high winds or 'gales' can bend the plants, thus soiling the cotton and reducing its value.

3.2 PRE-GINNING PROCESS

After picking, the seed cotton is loosely bundled into gunny bags, weighed on scale. The agents purchase this cotton from the farmers. If the ginning factories are in the vicinity, often the farmers sell seed cotton directly to the factories. The main operation carried out in ginning is to separate the cotton fibre from the seeds to which it is attached. For this, seed cotton must be presented to ginning operation in a proper condition to obtain satisfactory fibre quality. The work practices at ginning, storage of cotton and its transportation before and after ginning all influence fibre quality. The cotton fibre on seed with lower maturity tends to increase neps in ginning. The moisture in seed cotton also plays a vital role in ginning. Too moist cotton forms the wads and leads to chocking during ginning. On the other hand, too dry seed cotton poses problems of static. Also, when the cotton fibre becomes too dry, it loses strength. This results in fibre breakage during ginning. It, therefore, becomes important to control the relative humidity in ginning room to obtain consistency in the fibre quality of ginned cotton.

Many ginneries employ pre-cleaners before actual ginning to remove heavy impurities and immature boll fragments. Some of them even give post-ginning cleaning to remove fragmented trash particles from ginned cotton. This is practiced in most developed countries, especially when the cotton is machine picked. This enables very clean cotton to be subsequently packed. The pre- and post-ginning treatment have been attracting even Indian ginneries.

The operation, however, must be done with a lot of care so as not to damage the fibres. This is because, when the seed cotton is dried, it becomes easy to detach the fibres from seeds; but the fibre due to less moisture becomes weaker. Basically the machines employed may be divided into two classes: 'Cleaners' and 'Extractors'. The function of the cleaner is to open out the locks and lumps and to disintegrate the clods around the burrs and other trashy matter. The extractor, on the other hand, separates and removes the coarser foreign matter such as burrs, hulls, sticks, stem portions and stones, as well as fine impurities like leaf-bits, fine sand and dust. For this, a mesh is provided and it screens the impurities with the help of suction fan, especially the lighter ones. Some typical machines employed are as follows:

Seed Cotton Opener: The opener has two spiked rollers working with two sets of grid bars placed below (resembling to Axi-Flow). The material is fed by a feed lattice and the delivery of the material is received by another lattice.

Inclined Cleaner: It resembles Step Cleaner and has 4–6 spiked cylinders spaced equally at an inclined plane. A screen in the form of grid bars is provided under each cylinder. After the cotton is fed to the lowest cylinder, the material travels upwards and is finally delivered by the top-most cylinder.

Extractor: The machine contains saw-tooth cylinder and cleaning brushes. Feed regulation is also provided to feed rollers to ensure the quality of delivered material.

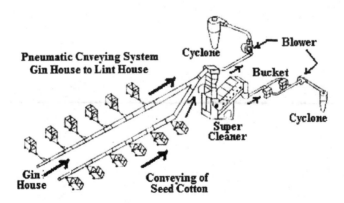

FIGURE 3.1 Pre-Ginning Line with Inclined Cleaner[4,5]: It ensures sufficiently dried cotton still on cotton seeds. It helps in snatching the fibres off the seed surface.

3.3   GINNING

After the pre-ginning treatment, seed cotton is ready for seed extraction. Saw Gin
and Roller Gin are the two popular methods; both can be employed for short and
medium staple varieties. Even though the ginning percentage is higher, the neps are
produced in Roller Gin. The Saw Gin, however, tends to damage the fibres owing to
forceful blows of saw teeth during detachment of the fibres. Macarthy Gin is another type of gin which is slow in its working and hence production rates are
comparatively much lower. However, the operation is more gentle and hence it can
be beneficially used for long staple cotton.

3.3.1   SAW GIN

The saw gin mainly consists of saw gin discs placed on a shaft. Each disc is approximately 12 inches in diameter and spaced at a distance ¾". A rib is a curved
plate with appropriate slots just enough for some portion of the saw teeth of each of
the discs to project through in the upward direction.

The seed cotton is fed by the feed lattice which is situated above the saw tooth
discs. The material before entering the disc zone is treated by a pair of guide rollers
and a spiked cylinder, both having spikes on them. The hopper receiving this
material finally leads the seed cotton to the actions of saw tooth discs. The teeth of
each saw tooth disc pick up small portion of fibrous matter attached to the seed,
pluck it off and take it through the rib slots. The seeds being greater in size than the
rib-slots are not able to follow this path and remain at the top of the rib. As the rib is
slanting downwards, when the fibres are detached by the sharp teeth, the seeds
eventually fall down and are collected in the seedbox situated below.

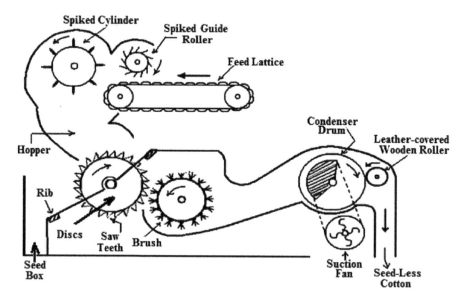

FIGURE 3.2   Saw Gin[2,4]: It uses saw tooth discs to snatch the fibres off the seed surface.

As mentioned earlier, the rib slots are just sufficient for the discs to freely rotate, and thus the tiny mass of the fibres plucked from the seed cotton is easily able to pass along with the saw tooth points of discs through this gap. The fibres are brushed off subsequently by a fast rotating brush having about 18 inches diameter and with its bristles touching the saw tooth points of discs. Finally, a strong stream of air is made to sweep the brush and carry loose fibrous lint on to the condenser. The suction is created inside the condenser and it helps to direct the flow of lint on its surface.

3.3.2 Air Blast-Saw Gin

In another type of saw gin, an **air blast** is used (as shown in Fig. 3.3). Here the air nozzle is placed in such a way that the blast of air is almost tangential to the tips of saw teeth. This effectively strips-off the fibres held by the tips. The fibres thus released are carried by a strong air current generated at the condenser. The damper inside the condenser cuts-off the suction at a certain position. Hereafter, a leather-covered roller, placed on the condenser helps the fibres to be taken from the condenser surface and deposits them into big trays put underneath.

3.3.3 Macarthy Gin

It is simple in construction and operation. As mentioned earlier, this type of gin works at much slower speeds and protects the fibre length. Hence, it may be beneficially used for longer staple varieties. The hopper situated behind carries the stock of seed cotton which is pushed forward by a reciprocating feed bar. The cotton thus pushed comes in contact with a leather-covered roller having rough surface.

The roughness of this roller makes the fibres on the seeds to get attached onto their surface. A fixed knife is set very close to this surface. This distance is just sufficient for the fibres to pass through. The friction between the roller and the fibres, thus, pulls the fibres through this gap between fixed knife and the surface of leather roller. This gap, however, does not permit the seeds to pass through. The

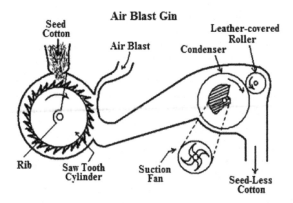

FIGURE 3.3 Air Blast Saw Gin[2,4]: It uses a powerful air-blast to strip off the snatched fibres from saw discs surface.

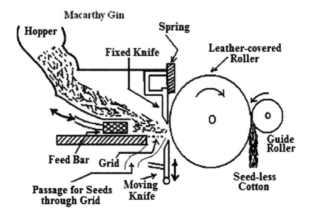

FIGURE 3.4 Macarthy Gin[1,2]: A slow speed machine used to protect longer fibre length.

frictional force between the fibre and leather roller surface overcomes the force with which the fibre are attached to the seeds and snatches the fibres from the seed-base. The fibres are then carried ahead around the leather roller.

If the frictional force is not able to snatch the fibres from their base, the seed remains caught at fixed knife. There is another moving knife which also reciprocates up and down. In its upward motion, it comes very close to and over the fixed knife.

The movement of this moving knife is matched with the reciprocating feed bar. When the bar recedes, the reciprocating knife is made to move up and ride over the fixed knife edge. It is at this time that this knife cuts the fibres very close to the base of the seeds caught at fixed knife. The fibres on the other side of fixed knife are still held in position by the frictional force between them and the rough surface of the leather-covered roller. Thus, when the seed is sliced by reciprocating knife, fibres on the other side become free off the seeds and are carried forward. They are subsequently deposited into the fibre tray. The seeds which get detached fall through the grid below and are separately collected in seed box.

3.4 POST-GINNING OPERATIONS

It is a common experience that after removing the seeds, the weight of actual lint cotton available is reduced to almost less than 40%. This means that a sizable proportion of seeds is thus available for subsequent operations. Even then, it is also a common practice to subject these seeds for second and third ginning operation (first-cut and second-cut linters) depending upon the need. The fibres extracted in this way are often termed 'linters' and their quantity is also sizable. These linters are either used domestically or exported. After extracting seeds from cotton, part of them is retained for sowing and the remaining is used for extracting lint-seed oil which has variety of end uses, one such being in the form of edible oil and soap industry. The remains after the extraction of oil are best cattle fodder and, therefore, often converted into cake form. Being very rich in their protein and fat content, these cakes are in great demands and are often purchased by the big dairies to improve milk yield from animals.

3.4.1 Care during Ginning

It may be mentioned that presenting too damp cotton to ginning makes the cotton stringy. This results in loose withdrawal of entangled strands. Excessive fibre cutting may also occur. This is more pronounced with long staple cotton when even 8% moisture level can make it damp. With short staple cotton, however, this limit is a little higher. In this case, even 11% moisture level may not make it damp. Excessive nep generation is very likely to occur when cotton is ginned in damp condition. The worn-out and blunt saw teeth aggravate the nep formation. The neps may be even produced when there is overfeeding of seed cotton to ginning machines. There are processes called 'carding' and 'combing' in the subsequent spinning operations, where the neps are more effectively removed. Even then, the neps are likely to persist thereafter. Especially with the fine fabrics, the presence of neps in the finished fabric mars the fabric appearance. The situation gets aggravated further when such fabric is dyed, particularly with the lighter shades. The neps pick-up the dye matter with different shades and this becomes more distinct (speckled appearance) when compared to the other portion of the dyed cloth.

Another danger is when the seeds get crushed or broken during the ginning process. Especially, the broken seed particles get mixed with lint. While removing these broken seed particles, more amount of cleaning becomes essential in the subsequent process. Further, owing to the severe blows of saw teeth, the crushing or even shattering of the seeds is likely to occur. This is still more serious, as it allows a certain amount of seed oil to ooze-out. This results in more damaging contamination of the lint. In both the cases, it lowers the value of the lint-cotton.

Even with the best-set machines, it is highly unlikely that all the fibres are taken out from the seeds in one single operation. Thus, a seed may be presented to ginning operations for more than two times for removing most of the fibres. However, the lint-fibre from first ginning is only useful for normal cotton spinning. This is because the length of the fibres obtained from first ginning is the best. The succeeding ginning operations only bring comparatively much shorter length fibres. The fibres obtained from these succeeding ginning operations are, therefore, used in either waste spinning industry, for low-grade yarns; or they become the cellulose raw material for rayon manufacturing or paper industry.

3.4.2 Trash before Packing into Bales

The seed cotton which is machine picked arrives at the ginning factories with trash content, sometimes in excess of even 25% in the the United States, Australia, Uzbekistan and other countries. However, when it leaves the baling factories in the form of pressed bales, it hardly contains 1% to 2% trash. Even in African countries like Uganda, Tanzania, Sudan and Egypt, where the cotton is hand-picked like in India, trash content in the bales is comparable with that of US cotton. This is because of good house-keeping and the use of pre- and post-cleaning machines in their ginning system.

In India, on the other hand, the hand-picked seed cotton arrives in ginning factories with substantially less trash than the machine-picked American or Australian cottons. However, when it leaves the ginning factory, the trash content in the Indian bale cotton is higher than that from these countries. Excessive quantities of foreign matter due to improper picking and ginning practices have earned notoriety for Indian bale cottons as the most unclean bale cotton in the world. In the recent past, some efforts have been started by the ginning/baling factories to improve this situation.

3.5   COTTON BALING[1,3]

Though in India, cotton is usually directly baled after ginning, in many foreign countries, some preliminary cleaning is given to extract trash from ginned cotton. The practice of post-cleaning is far more beneficial. Here, a considerable amount of trash can possibly be removed when the cotton is in loose form before being compressed. This reduces a lot of burden of opening-cleaning operations which are required to be carried at blow room in the mills. When the cotton is thus cleaned before packing into bales, it helps in increasing its value.

TABLE 3.1
Bale Density & Type of Bale[3]

Bale Type	Density Range Lbs/ ft³	
Flat Bale	15–20 (250–312 kg/m³)	In U.S.A. low density bales of 12–15 lbs/ ft³ (220–240 kg/m³)
Standard Bale	25–33 (416–521 kg/m³)	are produced for local market for a short distance transport.
High Density Bale	36–42 (580–660 kg/m³)	In India medium density bales are produced and the weight of the
Universal Bale	46 & above (729 kg/m³)	bales varies from 180 to 190 kg. In Egypt, very high density bales are produced with 350 kg/bale.

However, for long distance transportation and especially when the bales are to be shipped to different countries, heavy baling press is used to get a 'standard bale'. These bales have much higher density in order to reduce freight charges (they depend upon the volume). In spite of this, the weight of the bales remains more or less constant. In Egypt, both low and high density bales are made to suit the transport distance. In India and Pakistan, it is customary to produce high density bales of 550–600 kg/m³. The bale weight varies from 180 to 190 kg.

While compressing the cotton into a bale form, the jute cloth/net or cotton cloth is wrapped around to protect the inside cotton mass. The use of polyethylene slit-ribbons woven with jute is also becoming a popular choice. With polyethylene, the cotton is protected from getting moistened and soiled. The following types of packing material are often used to wrap the bales:

TABLE 3.2
Cotton Bale Packing Materials[4,5]

1. Woven Cotton Bags	2. Warp Knitted cotton Bags	3. Polyethylene Film Bags
4. Jute Bags	5. Polyethylene/Polypropylene Bags	6. Shrink Wrap

Finally in the compressed form, the iron bands are used over the packing material to tie the bales in position. These bands are quite thick, strong and sturdy and are spot welded.

TABLE 3.3
Approximate Bale Size, Dimensions and Density[3]

Cotton	Gr. Wt. -Kg	Dimensions H × B × L cm	Volume-m³ Cub. m	Density-kg/m³
Australia	232	125 × 62 × 112	0.86	263
Brazil	186	122 × 45 × 45	0.24	750
W. Indies	186	100 × 105 × 65	0.68	264
China	241	82 × 70 × 48	0.27	874
Egypt	343	130 × 72 × 60	0.56	594
India	180	125 × 50 × 45	0.28	635
Turkey	220	95 × 65 × 105	0.66	330
Kenya	186	102 × 68 × 68	0.47	382
Peru	218	108 × 70 × 108	0.81	264
Sudan	197	102 × 68 × 68	0.47	408
Russia	180	98 × 60 × 70	0.41	431
U.S.A.	227	150 × 80 × 50	0.60	371

LITERATURE REFERRED

1. Cotton Spinning – William Taggort.
2. Manual of Cotton Spinning – "Opening & Cleaning" – Vol II, Part II – W. A. Hunter & C. Shringley, The Textile Institute Manchester, Butterworths, 1963.
3. Bale Survey – 1995, International Advisory Committee.
4. Elements of Spinning – Dr. A. R. Khare, Sai Publication.
5. Blow Room – NCUTE Training Programme, I.I.T. Delhi.

4 Cotton through Blow Room

4.1 BLOW ROOM OPERATION[1,2] – WHY?

Many a time, textile mills are located far away from the cotton fields; whereas the ginning factories are somewhat closer to it. The ginning process removes the seeds from the cotton picked from the fields. After ginning, the cotton is in a loose form. It would, therefore, be highly uneconomical to bring the cotton in this form, as it is, to the mills. It would not only involve very high freight charges; but also storing this cotton in loose form would also involve very huge space in the mill premises. Thus, huge go-downs in the mills would occupy a very large space and it would be exposed to many hazards, e.g. fire catching and dampening. Hence, to facilitate economic transportation and to avoid the risk involved in storing loose material, the cotton from ginning factory has to be brought to the mills in the form of compact 'Bales'. This, however, puts a burden on the initial process employed in the mills – blow room which again is expected to loosen out this compact and hard-matted mass of cotton.

After the bolls burst open, they are exposed to the sunlight to allow the inside wet mass to get fully dried. During this period, the opened bolls are also exposed to the high winds blowing across the field. The winds, along with them, carry some dirt and sandy matters, which conveniently get settled on these open bolls. During the subsequent picking operation, especially if it is machine picking, it is equally likely that along with loose, hanging cotton, some vegetable matter is also picked up by the machines. Even during ginning, depending upon the machine conditions, certain fragments of seed coats and broken seeds accompany the ginned cotton. Finally, the journey of the bales from baling factories to the mills is equally likely to add some fine dust to settle on the bale surface if the wrapping cloth around the bales gets torn or becomes untidy. Hence, when the bales are brought to the mill godowns and later transported to the first processing department – blow room, the surrounding iron bands are first broken and the inside hard-matted mass is allowed to loosen on its own. The time is given for the bales to acquire the natural moisture regain level. Only after this, the blow room processing starts. The main object of blow room is to open and loosen the matted mass of bale cotton and to remove the foreign matter. Thus, the blow room tries to clean the cotton as far as possible.

Blow room is not a single machine operation, but there is a sequence of machines arranged one after the other to progressively carry out the job of 'opening' and 'cleaning'. Thus, the number of machines and their type in this sequence are generally governed by the condition and class of cotton to be processed. It is a common experience that American and Egyptian varieties of cotton are much cleaner than their corresponding Indian varieties. Hence, these varieties require a comparatively milder opening-cleaning treatment. Even the number of machines used in the blow room sequence is much less in such cases; whereas, many Indian cotton, being more trashy and dirty, require a stronger cleaning action. Obviously, it is expected that more number of machines in the blow room sequence are involved in carrying out the cleaning job.

In India, there has been a trend to spin multiple counts (from coarse-medium to fine-superfine) to diversify the products. The mills while producing coarser and cheaper yarns also tend to work fine and superfine counts to cater to manifold needs. Hence, the mixings that are worked in blow room also vary in their fibre properties, especially in their trash content. The blow room, therefore, needs to be more versatile and adaptable to handle the cotton mixings greatly varying in trash. The machines used in a blow room sequence, thus, need to be of both the types – those treating the cotton mildly and the others giving intensified opening/cleaning treatment. These machines are arranged one after the other, in a sequence, and hence while selecting them for a particular mixing; a correct choice has to be made. For this, there are by-passing arrangements suitably provided to some of those machines which give more intensified and harsher action. However, the machines which are more versatile in their action do not need such arrangements.

4.2 CONTAMINATIONS[7]

Apart from trash from vegetable, sand and fine dust, there are other impurities which also contaminate the bale cotton. These are in the form of alien fibres, coloured threads, cloth pieces, human hair, plastic films, paper bits, metallic objects and oil-grease, which are very often found in the bale cotton. They are mostly due to the negligence on the part of personnel who have the opportunity in handling the earlier operations (farmyard, market yard, ginneries and the transportation at every stage). It even includes the bale opening when they are brought from mill godowns to the blow room. These contaminants, especially in fine varieties of cotton, are not only source of trouble during processing; if they remain till the yarn production stage, they are bound to deteriorate the yarn quality. In modern blow room operations, there are many electronic devices which detect these contaminants earlier and remove them at the beginning of blow room processing.

It is well known that foreign matter content and impurities in Indian cotton are always on the higher side. The contribution of ginning factories to these is also quite significant. In the list of "most contaminated" cottons compiled by Indian Cotton Manufacturing Federation (ICMF) on the basis of extensive survey, some of the "best" Indian varieties occupied prime place.

Though the fibre properties of many types of Indian cotton are excellent, they are marred by excessive trash and contamination. The ginneries face difficulty in addressing these issues due to the seasonal nature of activity, which does not ensure consistent employment for the workers. Modernization through technological upgradation is the only solution to ginneries. Among various studies/surveys conducted by different organizations, Textiles Committee also conducted a countrywide techno-economic study of ginning and pressing factories. The study covered cotton seasons in two consecutive years and the report was submitted with few recommendations to the Government. Later on the basis of this report, TMC

(Technology Mission on Cotton) was set up. The main object of this mission was to improve productivity and quality of cotton. The TMC focused its attention on the modernization of farmyards, market yards and ginneries.

It is now realized that though the basic characteristics of Indian cotton are good enough to meet the challenges of yarn quality in terms of strength and evenness, they usually contain higher contaminants, especially the trash content in the bale cotton.

The following contaminants, when present, are considered serious.

1. Strings of jute or pieces of hessian cloth
2. Organic matter – leaves, feathers of birds, paper or leather pieces
3. Strings of pieces of plastics
4. Hair – other than cotton fibre (usually human)
5. Small pebbles
6. Oily or greasy rags

Some of the mills take precautionary measures by selecting the cotton in the field, carry out the ginning under strict supervision and see that the bales packed are transported to the mill godowns with hardly any addition of contaminants. In the absence of such a procedure, the mill has to employ labour to manually pick up these contaminants before sending the opened cotton for blow room processing. In modern blow room, there are special electronic gadgets like "Optiscan" which automatically detect and extract the contaminants. The manual picking of these contaminants is not possible when 'Auto Pluckers' do the job of opening the cotton directly from bales in modern blow room line.

Equally important is to educate the people working in various pre-blow room operations and those employed in blow room itself. This would improve housekeeping. The use of traditional hessian cloth to wrap the bales also requires rethinking.

4.3 MIXING OF COTTON[1]

It is known that cotton grown in different environmental conditions develops variations in the fibre characteristics. Also, the same variety of cotton grown at different places shows some fluctuations. Even the cotton grown in the same plantation shows variations in the four important properties – staple length, fineness-maturity, bundle strength and trash content. Thus, cotton grown in one part of plantation may materially differ from that grown in another part of the same plantation.

The greatest variation occurs when cotton is baled. This is because the lots from ginning section continue to come, one after the other, in the baling department. When these bales are received in the mill, there is obvious variation in cotton properties owing to each of the bales having received its own share of cotton, grown under different conditions. The cotton constituting a bale may have been cultivated under different methods of cultivation. Equally possible is that the cotton from different stations coming to ginning factory may also go into a bale. All this leads to substantial variation in the fibre properties even within a bale.

It is, therefore, a good practice to thoroughly blend (or mix) the cotton coming from different bales to represent a mixing. This ensures that the variations in fibre properties are spread more uniformly within a mixing lot. As a result, the yarn made from this lot is expected to exhibit more uniform and consistent yarn properties over an extended period. Hence, it is not only desirable but essential to mix several bales together and for this, many bales from the same mixing lots are simultaneously opened for processing in the blow room. Yet another factor which, many a time prevails, is economic consideration. The mill management, at times, may decide to use some of the bales which are slightly cheaper owing to inferior fibres properties. This definitely involves a little sacrifice on fibre properties. The proportion of such varieties, however, has to be judiciously restricted so that the overall quality of the mixing does not deteriorate markedly. Even then, using such a technique definitely brings down the cost of the mixing.

The lots thus selected for a mix, usually have somewhat identical properties, except for those bales which are intentionally chosen to be inferior, solely for economic purpose. Even then, very coarse and very fine cotton or very short and very long cotton are never blended together. This is because mixing of such widely differing cotton would pose peculiar processing problems, eventually leading to a very irregular yarn. Similarly, the trash content in the cotton should also not differ very widely; otherwise, this again is expected to bring-in typical processing problems. Especially in blow room, it would be very difficult to choose the sequence of the machines that would justify their action on cotton varying greatly in trash content.

The fineness and maturity of cotton in lots are other problems that should be taken into account while mixing the cotton. The finer types of cotton are more delicate and as such require comparatively milder treatment. Similarly, immature fibres cause serious processing problems. They usually lead to the formation of neps in blow room and in the subsequent processes. Therefore, while deciding the cotton lots for mixing, important characteristics like length and fineness, trash content and even colour of cotton should be considered with a due weightage to the price of cotton.

For taking a large number of bales, at a time, in the blow room, however, would necessitate a large space around the initial bale processing machines. In modern blow room line, with automatic bale openers, such a large space is provided at the time of installation of this machine. Hence it is possible to take, at a stretch, more than 60 bales for a mixing. However, with older or conventional set-up, this number is very much restricted to less than 20 bales at a time.

4.4   TRASH IN RAW COTTON[1,6]

The impurities in bale cotton consist mainly of sand, whole or broken seeds, fragments of stalks and leaves. These get intermingled with the fibres so intimately that it is often a big task to separate the fibres from them. The sand is comparatively heavy and as such has much less clinging power to hold on to the fibres. It is thus easy to separate it. So also, it is comparatively easy to remove whole seed than broken seed particles. The association of seeds or broken seed fragments is seen only after ginning. Undeveloped seeds (due to pest attack) and broken fragments are usually more rough and thorny. Being very small and lighter in nature, they are more difficult to remove. They cling to the cotton more tenaciously and survive through various beater actions. Ultimately in the scutcher and through cross-rolls unit of carding, they are ground to produce fuzzy motes.

The seeds or broken particles should be removed as early in the blow room processing as possible. This is because, if they get crushed, cotton is contaminated with seed oil. If the small broken seed fragments survive through spinning and weaving process, they cause yellow-brown staining in chemical processing, especially with alkaline treatment.

When analyzed, approximately 65% of the blow room droppings consist of fragments of leaves, bracts and stalk of cotton plant as their clinging power is comparatively less. Therefore, their presence in the final yarn indicates poor cleaning in the blow room. The other impurities in the droppings are mineral matter, soil and dust acquired from cotton field. The various beaters used in blow room and in subsequent carding process are expected to remove the greatest bulk of these impurities. Even then a small proportion of trash does remain in the card sliver. In modern blow room and card sequence cleaning efficiency levels permit only a very small fraction of this trash (less than 0.5 percent) to appear in the card sliver.

4.5   CONDITIONING AND MIXING METHODS[1,6]

In very old days, the mixing was traditionally done by arranging the stacks. Mixing bins were used for this purpose. As the bale cotton was in very matted condition, some preliminary opening was given to it and the loosened out cotton mass was laid down in the form of horizontal stacks. Each layer of the stack thus stood for a representative mixture of each of the different lots used for the mixing. These stacks were allowed to stand for certain length of time so that the cotton in layer would get acclimatized with the surrounding atmospheric conditions. It may be mentioned that for this, a certain level of humidity was maintained around the stacks. This allowed the cotton to regain its natural moisture level and in the process, get strengthened. It also allowed the fibres to recover from the stress under which they were packed in a bale form.

While using the cotton from the stacks for further processing, it was taken in vertical sections over the full height of the stacks. This allowed each section to represent all the layers constituting the stack, thus leading to a good mix. In another method, after a prior opening, the cotton was pneumatically transported to the big mixing rooms. The lots, during the pre-opening, were used one after the other. They were then deposited in the form of big heaps in the same room. The bin-room thus contained different heaps of cotton lots identified to represent all the constituents of the mixing. Again, the lots kept in open condition within the room were allowed to get conditioned to the mixing room atmosphere, which was maintained at standards humidity level. In old days, these mixing rooms were usually positioned on the first floor. A large hopper attached to conveying pipe and placed on the first floor finally reached these opened out lots to the first opening machine of blow room positioned at the ground floor. The labourers were employed to carry the opened out lots of cotton from mixing room to the funnel of the hopper. Through the funnel, the cotton was simply made to travel down the pipe by gravity.

As mentioned earlier, one of the objects of conditioning is to relieve the cotton fibres from great stress under which they were tightly packed in the bale form. In modern blow room line also, it is done by allowing cotton to remain in the bale stacking room for a short duration after the steel bands around the bales are opened. The short staple cottons are generally stronger, thicker and yet more trashy, while longer staple varieties are more delicate, fine and less trashy. Hence, with comparatively lower humidity levels, it is easy to open and clean shorter staple cotton so that there can be maximum extraction of trash. The cotton fibres are hygroscopic and gain in strength with higher humidity. With longer and finer types of cotton being less trashy and more delicate, a slightly higher level of humidity allows the fibres to strengthen themselves. They, thus, stand better to the subsequent blow room treatment without getting ruptured.

Both stack and bin mixing required a much huge space in the blow room and they were time consuming. Even labour requirement was comparatively more. However, they ensured a more uniform mixture of mixing lots. It used to help in evening out lot-to-lot variations in the fibre properties, thus leading to a more uniform yarn. With modern trends and machinery, these two methods have become almost obsolete. Especially with the introduction of automatic bale opening machines, where it is possible to lay the bales in large number, the homogeneity of blending different lots is quite high, and hence modern installations of blow room do not need these older mixing methods.

4.6   CLEANING POINTS – PRIMARY & SECONDARY CLEANING[1,6]

With older set-up of blow room line, the concept of opening and beating were two separate operations. The blow room machines were, therefore, divided as those mainly carrying out opening of cotton; whereas the others mainly focusing their attention on its cleaning. The conventional blow room machines were accordingly arranged in such a way that initial machines gave more stress on opening whereas those placed later were designed to act strongly on cotton to extract as much trash as possible.

It is, therefore, thought that they could be more appropriately classified as 'primary' and 'secondary' cleaning points. This is because; any opening action involves some cleaning and therefore, initial machines that stress on more opening also extract some trash. It will be appreciated here that the secondary cleaning points are also important as they prepare the material and make it ready for primary cleaning points to do a more thorough job. Thus, in a conventional blow room set-up, a secondary cleaning point followed by primary cleaning point becomes a more complete unit for carrying out cleaning.

The primary cleaning points in conventional blow room are beaters like – Crighten Opener, Porcupine Opener, Bladed Beaters and Step Cleaner, while machines like Hopper Bale Breaker, Hopper Feeder, Blenders, Condenser or Dust Cages can be regarded as secondary cleaning points. The secondary cleaning points (except condenser/cages) give a light teasing and opening action on the matted lumps of cottons, reduce them in size and prepare them for a more vigorous action by primary cleaning points. It may be mentioned here that more number of machines than necessary does not necessarily mean more cleaning; on the contrary, they make the blow room laps stringy. The overprocessing always leads to weaken and damage the fibres and is one of the reasons for excessive nep generation.

The approach with which modern blow room line tackles this problem is much different. It does not separate opening and cleaning as two distinct operations. This is because cotton cannot be cleaned effectively and efficiently unless it is opened out. Further it has been now realized that these two operations must be done simultaneously; otherwise, there is a chance that lint and trash separated elsewhere, combine again (recombination of lint and trash). The long suction pipes conveying cotton from one machine to the other in blow room line often lead to this recombination. In modern blow room line these conveying distances, therefore, are comparatively much shortened.

Another important aspect is that each of the machines in blow room sequence must positively contribute to progressive opening of the cotton tufts. In other words, each machine in succession, must reduce the tuft size (mass), and thus bring larger surface area of the cotton tufts in contact with cleaning contrivances. It can be safely assumed that the cleaning can be far more effective from the surface of the cotton. Hence, when the tufts are broken to still smaller tufts, they expose larger area to the cleaning action of the machines, thus making it more effective. A continued tuft reduction (both in mass and size) at every stage of processing in blow room becomes the basic principle of any modern blow room line. It has also helped in reducing the number of machines in modern blow room sequence; and yet, giving comparatively higher cleaning efficiency without harshness of the treatment.

4.7   OPENING AND CLEANING IN BLOW ROOM[1,6]

After the bales are received in the mill, they are stored in the warehouses and, on an average, a minimum stock of 3–4 months is maintained. The purchase policy is usually decided by the central management. The technocrats working in the mills rarely have a say in this purchase. In some cases, the lots to be taken are also specified mixing wise, whereas in only some instances the choice is left with the senior technical personnel working in the spinning department. Thus, the lots specified for a particular mixing may contain a particular variety coming from different place (stations) and/or may also contain other varieties matching in fibre properties. As mentioned earlier, there could be some lots purchased purely on economic grounds and they are invariably poorer in one or more fibre properties. A technologist has to judiciously select the proportions of these lots constituting a mixing, so that the average properties of the mixing are maintained fairly constant over a sufficiently longer period. This may also ensure fewer processing troubles.

Once the proportions are decided, the bales are brought on a regular basis from mill godowns to the blow room department by hand-carts or motor-driven trolleys. With some extra space available in blow room, the quota of bales for the day may be brought all together. In very old days the blow room was divided into two or three compartments – mixing and mixing rooms, blow room preliminary sequence and finishing sequence.

In conventional blow room, it has been possible to amalgamate these sections into one single line called 'single process' blow room line. Thus, after removing the iron bands and hesian cloth wrapping, the bales are conditioned for a very short duration and are laid around the first machines in blow room sequence – Hopper Bale Breaker. Usually, there are two such machines or else, when Blenders are used, there can be 4–5 such machines. When Blenders are used, there is always an additional one provided for processing soft waste (A good reusable waste in subsequent spinning process, e.g. sliver and lap pieces).

In modern blow room, the number of machines is drastically reduced and their functioning style has been totally changed. The first machine, as mentioned earlier, is automatic bale opener around which many bales (usually not less than 60) are laid down for running a mixing.

4.7.1 HOPPER BALE BREAKER[1,6]

In old, conventional blow room, Hopper Bale Breaker assumed a lot of importance. It is, as the name suggests, the first machine in the conventional set-up that is used to tackle the hard-matted mass of bale cotton. The machine, therefore, has been made quite sturdy, strong and robust in its construction.

The long creeper lattice (Fig. 4.1) receives the lumps of cotton from the bales. All the bales are laid around this lattice. A tenter is instructed to take smaller slabs of cotton from each of the bales in turn. The creeper lattice moves very slowly and subsequently delivers the material onto the feed lattice which is situated at the base of a large hopper or bin.

The feeding is adjusted in such a way that the hopper formed above this lattice always remains about three-quarter filled with cotton. The feed lattice brings the cotton slabs very close to a faster-moving inclined spiked lattice. The spikes on this lattice pick up the small tufts from the slabs of the compressed mass of cotton brought ahead by feed lattice and carry them upwards.

Almost at the top, there is a roller covered with similar set of spikes and is called Evener Roller. The direction of the rotation of this roller is opposite to that of the spiked lattice. So also, the spikes of this roller are set very close to those of spiked lattice. This sets in a slicing action.

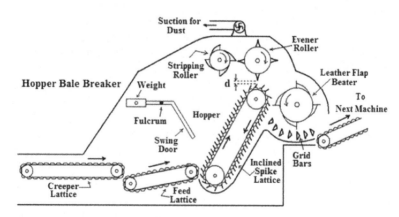

FIGURE 4.1 Passage of cotton through Hopper Bale Breaker[1,6]: The machine used to be placed in the beginning to open-out compact and matted mass of Bale Cotton.

The spikes of evener roller slice any extra mass of cotton that is carried by the spikes of inclined lattice. This results in great reduction in the tuft size. The tufts thus reduced in size are carried further and around by the spiked lattice. The distance 'd' between the spikes of evener roller and inclined lattice is adjustable. The narrower distance allows only smaller tufts to be retained by inclined lattice. It may be mentioned here that the rate of production of this machine largely depends upon this distance. The wider setting allows bigger tufts to pass through. However, this reduces the opening action. Therefore, for increasing production rate of this machine, the tuft-opening action at evener roller has to be sacrificed.

The excess of cotton sliced by the evener roller is returned back to the hopper. It may be possible that some of the tufts do not simply fall back into the hopper as they are retained by evener spikes. For this, there is another roller called stripping roller, which is made to strips the evener spikes continuously. The flaps of stripping roller ensure that the cotton thus swept falls back into the hopper.

With slicing action of evener spikes, when excess material is sliced-off the spikes of inclined lattice, some material is still withheld by inclined lattice spikes. The spikes on the lattice take this material which is appreciably reduced in tuft size, ahead through the gap between evener roller and spiked lattice. So also, it is equally likely that some of the tufts originally picked up by the spiked lattice are very small. They pass through the gap unharmed. In either case, the tufts which ultimately pass through the gap are finally stripped-off by another beater called leather flap beater. The speed with which this beater swipes the cotton tufts from the spikes of inclined lattice gives a whipping action to the cotton. The cotton tufts are not only removed from the spikes but, in the process, are also thrown over the grid bars. The mere force with which the cotton is struck against these grid bars forces some of the released impurities to pass through the small openings of grid bars. The cotton fibre-mass, being bigger in volume, is not allowed to go through these gaps so very easily.

The swing door is a panel door which is fitted inside the hopper and it extends across the whole width of the machine. This door is counter-balanced by a dead weight which can be adjusted for its leverage from its fulcrum. When this leverage is increased (weight shifted away from the fulcrum), more cotton is allowed to pile-up inside the hopper. As mentioned, the hopper is not allowed to gather material more than about ¾ of its capacity. When this limit is reached, i.e. when the cotton starts exceeding this capacity, the swing door is pushed against the moment of the weight. This is electrically conveyed to disengage the driving of feed lattice, and it stops bringing more material close to inclined lattice. It may be noted that more cotton gathering within the hopper has a tendency to roll the cotton mass, thus increasing the possibility of creating more neps.

The bale breaker being the first machine to tackle the hard-matted mass of cotton, there is always a considerable strain on the spikes of inclined lattice. This is because they have to carry quite big lumps of cotton. Therefore, it is always a good practice to instruct and train the worker to pick up as small and thin slabs as possible. In turn, this also allows a better representation of each of the bales in the mixture. Thus, it not only improves the blending intimacy of the mixture but also puts much less strain on the spikes of inclined lattice. There is, however, a tendency often shown by the worker to finish the job hurriedly. He puts the bigger slabs on the creeper lattice so that he can snatch some leisure time. The technical staff has to supervise and stop this very thing from happening.

Even then, it is still necessary that the spikes of inclined lattice are made stronger. For this, they are made of steel with about 6.4 mm diameter and tapering to a point. The density of the spikes on inclined lattice is about 190–195 spikes per square meter. Similarly, the spikes on evener roller are also made strong. However, they are a little longer in their length than the spikes of inclined lattice.

A suction fan is placed above the action zone of evener roller and stripping roller. Normally, owing to the action between the spikes of evener roller and inclined spike lattice, so also due to the stripping action of stripping roller, there is a lot of fly and fluff generated around this region. The fan rotating at quite a high speed helps in sucking away the flying, floating short fibres and dust. The sucked-out material is carried away and is finally deposited into cellar or filter unit. As mentioned earlier, the production of this machine can be varied by altering the setting between evener roller and inclined lattice. However, a better way to increase the production is to speed-up inclined spike lattice. In this case, the quality does not suffer. For normal settings, the production varies from 3300 to 3500 kg per shift.

4.7.2 Hopper Feeder[1,6]

The tufts delivered by the bale breaker are much reduced both in size and the mass. The action of the Hopper Feeder (Fig. 4.2), resembling in many respects to bale breaker in its construction, continues further the action of opening the tufts to still finer size. However, in addition to this, it has to play an important role. Similar to Bale Breaker, the cotton here is received on the slow-moving feed lattice, which, in turn, feeds it to the inclined spike lattice.

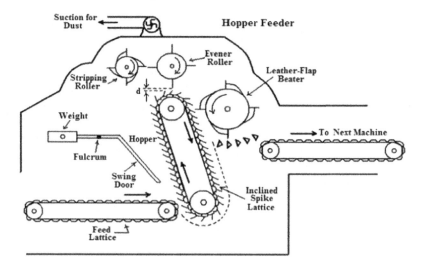

FIGURE 4.2 Hopper Feeder[1,6]: The basic function is to continue the action of opening of bale cotton and to control feed to the next machine in sequence.

The spikes on this lattice pluck the tufts from the mass fed by feed lattice and carry fine lumps of cotton tufts upwards. The evener roller with a closer setting to the spikes of inclined lattice further reduces the mass and excess of cotton carried by the spikes of inclined lattice is sent back to the hopper. The small cotton tufts that are still held by the inclined spikes, however, are carried further and around and are finally swept by the leather flap beater.

Here too, the action of leather flap beater is not merely of sweeping nature. The beater-flaps throw the cotton tufts on to the grid bars placed below the beater. Therefore, some cleaning is again evident here. Finally, the beater passes the cotton on to the delivery lattice, which carries it to the next machine.

The stripping roller does the similar function of cleaning the spikes of evener roller and the material thus stripped is sent back into the hopper. The exhaust fan situated above sucks off the finer impurities and floating fluff and carries them into waste cellars or filter bags.

4.7.2.1   Comparison of Bale Breaker & Hopper Feeder

As compared to Bale Breaker, Hopper Feeder is not very powerful and strong in its action and sturdy in its construction. This is very obvious as the cotton already reduced in tuft size in Bale Breaker is worked on by Hopper Feeder. The spikes on inclined lattice are, therefore, finer and their point density is much higher (648 points per sq. metre as against 240 points per sq. metre with bale breaker.). The finer spikes thus help in picking up still smaller cotton tufts and carry a great opening action.

The angle of inclination of the inclined lattice is also different. In Bale Breaker, it is obtuse, as much matted and heavy lumps from bale slabs are required to be lifted up. As against this, the acute angle in Hopper Feeder ensures that no bigger tufts are lifted up. The angle itself makes it difficult for such heavy tufts to be lifted up so easily.

In place of evener roller, in some Hopper Feeders, there is evener lattice (Fig. 4.3) used which offers more spike-points per unit area to assist better opening action between inclined lattice and evener lattice. In such a case, the stripping roller is not used. The evener lattice, in place of evener roller, offers more surface area and provides more spike-points to chop-off the excess size of cotton tufts carried by inclined lattice, thus assisting in better opening action.

In short, Hopper Feeder, as a modified version of Bale Breaker, is meant to give more opening to the cotton tufts. There is another version of Hopper Feeder called 'Hopper Opener' (Fig. 4.4).

But the spikes of inclined lattice, in Hopper Feeder are much finer and highly dense as compared to those in Hopper Bale Breaker or even Hopper Opener. This helps in further improving their opening action on cotton tufts and thus helps in reducing the tuft mass and size. Yet another point in favour of Hopper Feeder or Opener is that the construction of the leather flap beater. Along with the leather flaps, the beater also has spikes. Hence, the beater is able to give very effective opening and striking along with flapping action.

Basically, Hopper Opener is used as an extension of bale breaking action. Its main function is to carry further breaking and opening of the tufts and continue the

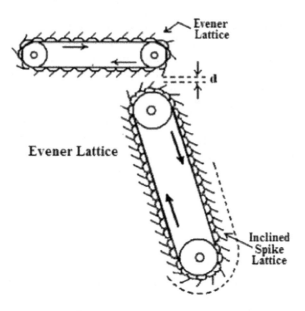

FIGURE 4.3 Evener Lattice[6]: It evens-out the material flow by restricting the tuft size passing through.

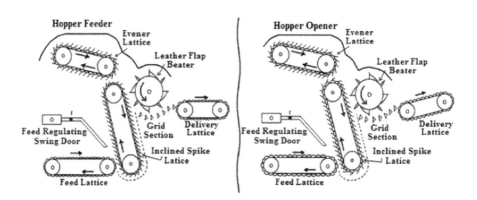

FIGURE 4.4 Hopper Opener & Hopper Feeder[6]: They almost look alike and are basically meant to continue opening of the matted mass.

action from where Bale Breaker has left. Hopper Feeder has not only to see that the cotton lumps are broken down further to very small tufts, but also to give a controlled feeding.

Hence, Hopper Feeder is used at different positions in the sequence of blow room to carry this important task – 'controlled and regulated feeding' to important beaters in the sequence.

4.7.2.2 Principles of Feed Regulation

It is very important that every machine in the blow room sequence must be supplied with a regulated and controlled amount of feed. When this feed is in excess, the machine cleaning capacity gets seriously impaired. It is, thus, essential that each machine is fed with a constant and correct flow of quantity of material so that maximum opening and cleaning can be achieved. In reserve box, (Fig. 4.5) the cotton is allowed to pile up inside the box to a certain level (height). This ensures a constant quantity of cotton all the time within the box. This is achieved by a control lever.

This lever is fulcrumed at (A_2) and has a weighted arm (A_1) on one side and a light panel door (A_3) on the other side. The panel door (A_3) projects inside the reserve box. The weighted arm balances the weight of the light panel door. When the cotton falls into the reserve box and reaches a certain height above the panel door, the weight of the cotton presses the door (A_3) downwards, thus raising the weighted arm (A_1). The resulting clockwise moment of arm (A_4) disconnects the circuit through a micro-switch. Eventually, it stops the incoming feed from the previous machine into the reserve box. At the bottom of the reserve box, there are two delivery rollers which are made to rotate and thus deliver the material piled-up within reserve box. This reduces the cotton height in the reserve box and again allows the panel door (A_3) to rise up owing to the moment of weighted arm (A_1). The arm (A_4) swings back in anti-clockwise direction. The signal through micro-switch again restarts the feed into the reserve box. Thus, a constant level of cotton within the box is continuously maintained.

It is very necessary to have cotton piled up to a constant height. This is because, the cotton being loose enough, tries to settle down slowly when the machine stops for a certain length of time.

The density of the cotton at the bottom delivery rollers, thus, is likely to fluctuate when the height in the reserve box constantly varies. This is very likely to result in the variation in the mass of cotton delivered by the bottom delivery rollers on to the delivery lattice below. Even when the machine works continuously, the varying mass of cotton,

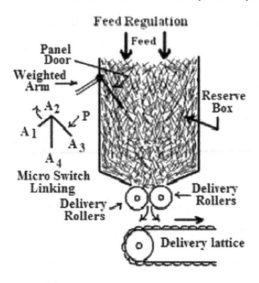

FIGURE 4.5   Reserve Box[1,6]: For regulating the feed to the subsequent machine.

due to incorrect maintenance of cotton level in the box, changes the density of outgoing cotton. Therefore, the use of reserve box equipped with the control lever efficiently regulates the amount of cotton delivered to the next machine.

Another method of controlling the level of cotton in reserve box is by using photo-cell (A – Fig. 4.6) along with a light source (B). As shown in the figure, the photo-cell and light source are placed within the box exactly opposite to each other. Thus, the light rays are made to fall parallel on to the photo-cell. As long as the light falls uninterrupted on photo-cell, the current generated relays the signals to continue feeding into the box.

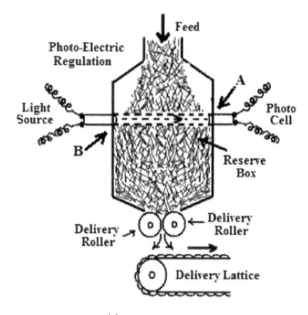

FIGURE 4.6   Photo-Cell Principle[1,6]: It is based on electronic principle. When the light is interrupted, the feed stops.

When the material reaches above the light path, the light falling on the photo-cell is interrupted and the signal stops the feeding. In this way a constant height of level of cotton within the box is maintained. Some machinery makers follow yet another simple method of maintaining height, especially at scutcher. The reserve box is slightly overfed to pile-up the cotton in it while a wiper roller is provided to wipe-off surplus cotton back into the hopper which is behind the box and which supplies the cotton to the reserve box.

4.7.2.3   Evener Roller Setting

As mentioned earlier, the setting between evener roller (or evener lattice) and spikes of inclined lattice, in Bale Breaker, Bale Opener or even Hopper Feeder affects the degree of opening of cotton mass, It also regulates the flow of cotton going ahead, thus controlling the production rate of the machine. The closer setting between these two reduces

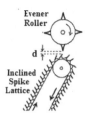

FIGURE 4.7 Evener Roller Setting: The setting decides the quality of work done and allows only specific size of cotton tufts.

the size of the cotton tufts that are allowed to pass through. This gives a greater degree of opening. As a large amount of cotton is returned back into hopper, and comparatively less cotton is allowed to pass ahead, the rate of machine production is reduced.

While arranging the blow room machinery sequence, it is necessary that there is a perfect balance of production between each machine so that all the machines in the sequence run continuously to give maximum machine utilization. In case, the closer setting reduces the production of a hopper feeder placed in the line, the machines ahead is likely to be underfed and would stop intermittently for want of cotton supply. Therefore, it is always advisable to slightly over-feed the cotton to the next machine where the feed control mechanism is attached. This ensures that there is a certain minimum supply of cotton that is required to maintain a continuous flow of the material. This always avoids any shortage of cotton supply to the next machine/s subsequent to the Hopper Feeder. A controlled and continuous flow also ensures that the beaters work at their best and give maximum cleaning efficiency. In this connection, a closer setting may be accompanied by increased speed of inclined spike lattice. This, while assuring a certain degree of opening, would also provide the required production rate of Bale Breaker or Feeder.

4.7.3 AUTOMATIC HOPPER FEEDER[1,6]

When equipped with feed control systems, the Hopper Feeder can be used at any place in the blow room line, especially prior to any machine where uniform feeding is required. A criticism is always levelled against the excessive use of this machine as it leads to curling of fibres and develops stringiness due to rolling and tumbling action of both, inclined spike lattice and evener roller.

The incoming feed to this machine (Fig. 4.8) is through Shirley wheel. The cotton is dropped into reserve box. As mentioned earlier, the reserve box feeds the material onto the feed lattice through a pair of delivery rollers. Thus, a controlled amount of feed is ensured on the feed lattice. The action within the hopper feeder is the same and cotton picked up by the inclined lattice is worked by the evener roller which chops-off larger tufts and sends them into the main hopper. The mass of cotton in the hopper is also controlled and when there is an excess of cotton gathered in it, the supply from feed lattice and delivery rollers is stopped.

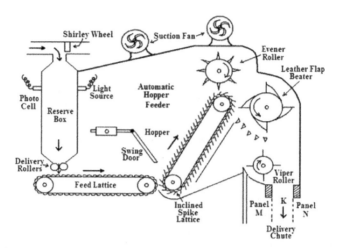

FIGURE 4.8 Automatic Hopper Feeder[1,6]: It is equipped with both mechanical and electronic contrivances to control the feed into machine, through machine and going out of machine.

The indication of excess material within the hopper is sensed by the swing door arrangement. When it is pushed down owing to the pressure of excess cotton mass within the hopper, a signal is relayed to stop the feed lattice.

The volume of the cotton in the main hopper is mainly decided by the amount of cotton carried by the inclined lattice and the one taken ahead to leather flap beater. It is also governed by the spike density on the inclined lattice and the pressure developed by the movement of feed lattice in bringing the incoming cotton closer to the inclined spike lattice. Even when all the spikes are supposed to pick up cotton tufts, some of them move without any material, whereas the others pick up varying sizes of tufts. However, when this material is brought under the action of evener roller (or evener lattice), only those tufts which are either smaller than evener roller setting or which are oversized and chopped-off, successfully go ahead for subsequent action of leather flap beater. The size of the tufts there onwards is uniformly small.

The material is finally delivered into a chute (K). The arrangement here is such that the chute (K) is made to slightly overflow. The excess of cotton, however, is deflected back on to spike lattice by the Wiper Roller (L) and this cotton is carried back by the lattice to main hopper. The cotton in the delivery chute is always maintained at a constant height. The two panels – (M and N), are made to continuously vibrate and oscillate. This makes the cotton in the chute more compact, thus improving the material regularity.

The height of this chute is about 10–12 feet and the mass of the cotton in it, at any time, is approximately equal to the weight per unit length of the lap being formed. The panel sheet N is adjustable and the gap between M and N when widened also increases the lap weight per unit length. The sheet below N is made transparent to enable the operative to observe the level of cotton within the chute.

As mentioned earlier, the Hopper Feeder is basically used anywhere in the blow room sequence, but more so in the later part of the machinery sequence, especially prior to scutcher. This is because; it is the scutcher where the final blow room lap is formed. As seen, the Automatic Hopper Feeder controls the material in three ways: (1) Material brought to it by feed lattice (2) Material within the hopper and (3) Outgoing material through chute. The oscillating sheets or panels – M and N make the sheet of cotton more compact and uniform and this helps the final lap making process. However, when the hopper feeder is used earlier in the sequence, these panels are not necessary as they prematurely make the cotton sheet more compact and impede the process of opening and cleaning in the subsequent machines.

4.8   BEATERS & CLEANERS[1,6]

Bale Breakers and Hopper Feeders are the preliminary machines, basically used for breaking open the matted mass of cotton; however, they extract a little amount of trash. They merely open out the cotton lumps into smaller tufts and in doing so, whatever trash or impurities that are liberated free, are released through the flap beater-grid bar action. For more thorough cleaning, however, several other beaters and cleaners are necessary to perform good efficient cleaning action. Owing to their higher speeds, the beater blades fling the cotton more forcefully on the sharply edged triangular grid bars. This leads to very effective opening and cleaning. Incidentally, these bars are spaced with a comparatively narrow gap in between them. The mere force of the beater blades flings the cotton on to the sharp edges of grid bars, thus releasing sizable foreign matter or trash. The trash being denser and much smaller in its size easily passes through the narrow gaps between the grid bars; however, the opened out cotton which is bigger in volume cannot.

With varying construction of different beaters, only the size and shapes of the striking element differs. Even then, the principle of using these striking elements against the grid bars remains more or less the same. The intensity of the striking force depends upon the beater speed, the shape and the size of the striking elements. Depending upon this force, therefore, the liberated impurities are forced through the grid bar gaps. In old traditional blow room, different types of beaters were used to carry out the job of cleaning. The selection of such machines largely depended upon whether the cotton was dirty-trashy or comparatively clean.

In modern blow room, the concept of cleaning has been radically changed. The modern blow room sequence basically involves reduction of tuft size and mass at every stage. This is because it is known that reducing the tuft size exposes a larger surface area. It can be safely assumed that the cleaning of cotton basically takes place from its surface. So, breaking down the cotton tufts into smaller size automatically exposes more surface area from which it becomes easy to clean. Now, it is also known that any machine which does not continue its action of breaking down the cotton tufts can be considered to be redundant. This is because, under the situation, the machine can hardly be expected to carry out cleaning job.

Thus, it is evident that both cleaning and opening have to go hand in hand. These have to be carried out simultaneously. This is because cotton can only be cleaned when opened sufficiently. In this respect, the action of Bale Breaker or Hopper Feeder and other cleaners or beaters differs materially. Some common beaters or cleaners are described in the following section.

4.8.1 PORCUPINE OPENER[1,6]

This machine has been extensively used in conventional blow room line as preliminary opener. It also does a great job in cleaning the cotton. Apart from being a good cleaner, the opening action of this machine has been really excellent. Therefore, the feed from this machine is normally directed to one of the blow room machines carrying out intensive cleaning. This has increased its popularity and in the past, every blow room line used to include this machine in its machine sequence. Owing to its powerful opening followed by effective cleaning, this opener is specially used when processing low grade cotton varieties, containing relatively high percentage of leaves that are normally difficult to separate from cotton fibres. Owing to its powerful opening action, porcupine opener is positioned early in the blow room line.

The cotton is received by the lattice in the fleece form (Fig. 4.9). The pair of rollers (A) and (B) guides the cotton to a feed system of pedals (N) and a spring loaded pedal roller.

Another feed system of pair of feed rollers (C) finally feeds the material in the sheet or fleece form to a fast-moving porcupine beater. The overall diameter of cylinder is about 24 inches and has 18–20 circular discs along its width (in figure only one such disc is shown). These discs have sliced rectangular blades riveted near their circumference. As shown in Fig. 4.10, these blades are riveted on both the sides of each disc and are off-set at various angles. Thus, each blade on the disc being curved to a different angle, the total number of blades on each disc covers approximately 2 inches width on either side.

There are about 20 such discs across the entire width of the machine. Thus, the full width of the cotton sheet is under the striking range of all the blades across the machine.

This enables the porcupine cylinder to strike on the full width of fleece presented to it and does not allow any portion of this fleece to escape from the striking action of the beater blades. Further, as each blade is much thinner, the amount of cotton drawn ahead by each of the blade is much smaller, thus inducing very good opening action. On the underside of the beater, there are grid bars (Fig. 4.11) which are adjustable to the porcupine opener. An external lever connected with the bars can be moved over an engraved scale which indicates both the corresponding position of the bars and clearance between the bars and the opener.

Thus, the grid bars can be fully opened to widen the gap between them or completely closed to minimize this gap. Any intermediate position also can be obtained with the help of the external lever. This setting is governed by the trash in the cotton. A more open setting (grid bars open) allows more droppings through the grid bars. The droppings in this case usually would contain both the trash as well as some proportion of lint.

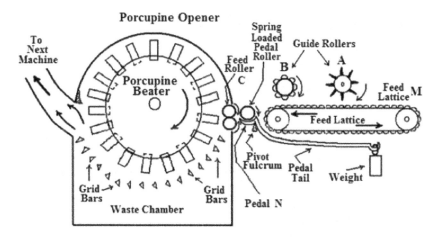

FIGURE 4.9 Porcupine Opener[1,6]: A typical machine with blades appearing as porcupine pointed arrows. It is one of the very powerful cleaners.

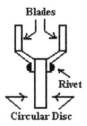

FIGURE 4.10 Beater Blades[6]: Off-set blades riveted to the central discs increase the opening action.

By closing the grid bars, the distance between each adjoining bars is narrowed down. This allows comparatively less trash dropping down and also minimizes the lint loss. Therefore, the setting of grid bars is required to be judiciously arrived at, so as to have a correct balance between the proportion of the trash and the lint in the droppings. The waste chamber (Fig. 4.9) which is fitted below the grid bar section, is equipped with an inspection window for observing the nature of the droppings. The window is also illuminated from inside so that by seeing the composition of

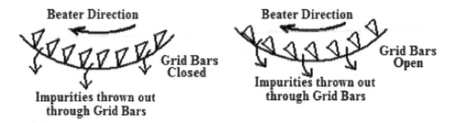

FIGURE 4.11 Angle of Grid Bars[1,6]: Within the reasonable limits, when the grid bar angle is changed, the nature of the droppings under the beater also changes.

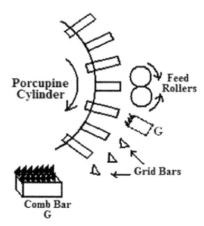

FIGURE 4.12 Comb[6]: A powerful contrivance to effectively open-out the matted mass of cotton tufts.

trash, the grid bar setting can be changed (In modern machines this can be done during the normal running of the machine).

Sometimes a transverse comb bar G (Fig. 4.12) is mounted. The comb is closely set with the blades. This bar is positioned immediately below the feed rollers. The teeth of the comb bar are self- stripping metallic wires with their sharp points facing the direction of the rotation of the cylinder blades. This greatly helps in reducing the size of the cotton tufts thus further intensifying the opening action. Eventually, with this powerful opening, when the tufts subsequently strike against the grid bars, a very effective cleaning follows. The beater usually runs at 750 r.p.m. The higher speeds may be used for shorter and coarser types of cotton though. A closer distance between pedal feed system and the opener blades is possible, whereas the minimum close distance from the opener blades is restricted to half the feed roller diameter in the case of pair of feed rollers. Therefore, with short staple cotton, the feeding system consisting of pedal and pedal roller needs to be used, whereas with longer staple cotton, a pair of feed rollers is preferred. Further, there is more of snatching with a pair of feed rollers and this defect allows bigger tufts, at a time, to be pulled from the feed system. This greatly reduces both opening and cleaning action.

With low grade and more trashy cottons, it was customary in the past to employ two Porcupine Openers in tandem i.e. two openers, one after the other (Fig. 4.13). When used in tandem, the openers are separated with a cage system interposed in between them. This allows the cotton to be conveyed from one porcupine to the next in the sequence.

The first opener in this case may be slightly bigger in diameter and may run at a slightly slower speed than the second one. By-passing arrangement is provided so that when two openers in succession are not required, the cotton material after passing through the first can be diverted to the next machine in the sequence. This is done by appropriately positioning the flap (F). The production of the machine is 500–550 kg/hr for 41-inch wide machine.

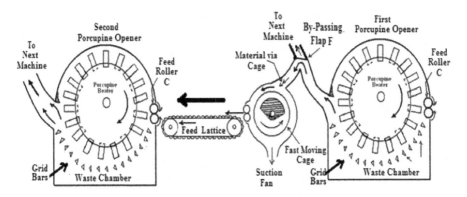

FIGURE 4.13 Porcupine in Tandem[1,6]: Two Porcupine Openers used one after the other, significantly improve cleaning efficiency of the blow room.

4.8.2 Crighten Opener[1,6]

This opener is largely employed for very low-grade cotton, which contained very high proportion of trash. This is because this machine gives the strongest and the harshest possible action on the cotton. Hence, for fine and superfine cotton varieties which invariably contained comparatively much lower trash proportion, it was never employed. Even in conventional blow room line, especially when processing cotton with moderate maturity, it was always discarded. However, with very coarse mixings (12^s–14^s or below), the trash content with some typical types of Indian cotton used to be 8% and above. In such cases, the Crighten opener was found to be very useful and hence was used in the blow room sequence.

The main feature of this machine (Fig. 4.14) is an inverted conical beater, which is placed in the vertical plane. There is a central vertical shaft (A) carrying six to eight circular discs (B). These discs are mounted on the shaft with cast iron spacing bushes separating each one.

Each disc carries a number of striker blades (D) around its circumference. Each of the striker blades is riveted to the circumference of the disc and is bent to differing angles. This makes the tip of each of these blades to follow a certain different path when the beater rotates. In fact, when the path of the tips is followed, it almost traces the threads of the spiral worm. With this arrangement, the striker blades of lower disc almost lead the cotton tufts within the beater to those of succeeding higher disc.

The diameter at the bottom including the blades is 18 inches while that at the top is 34 inches, with a gradual increase from bottom to top. The beater shaft with the beater is supported on anti-friction foot-step bearing which combines a thrust bearing and self-adjusting radial bearing.

At the top, the beater is supported with another radial type of bearing. The beater is surrounded by a fixed set of conical grid bars (E) and is provided with an arrangement (F) to raise or lower the whole beater. As the shape of the beater is inverted cone, raising or lowering the beater accordingly increases or decreases the setting of striker blades with grid bars respectively.

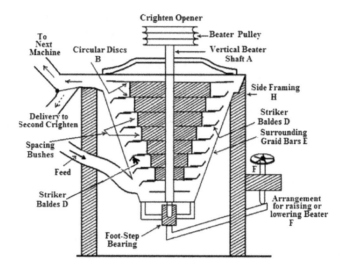

FIGURE 4.14 Crighten Opener[1,6]: A harsh beating beater. The cotton travels through the machine against gravity. Therefore, unless it is sufficiently opened to very fine size tufts, it does not rise so very easily.

As the grid bars have to extend a full distance from top to bottom portion of the beater, they are wider at the top and narrower at the bottom. With this arrangement, the beater is uniformly covered. It, thus, enables a uniform bar-to-bar spacing (gap) from top to bottom. The number of grid bars can be varied from 148 to 208 (in steps of 20 bars), the higher number being used for trashy, dirty and low-grade cotton. In some typical beaters, an arrangement is also provided for varying the spacing between the bars and angle of bars (Fig. 4.15).

This permits change in spacing between the bars. More open spacing between the bars and the sharper angle with which they face the oncoming cotton allows the trash to be easily ejected. This is required for low grade mixing. Obviously, for higher grade mixing, the gaps are required to be narrowed and their angle needs to be less sharp. Some machinery makers also provide perforated screen surrounding the grid bars.

4.8.2.1 Beater Functioning

The cotton is fed into the beater chamber by a feed-pipe, and it enters the casing at the lowest position. Immediately the striker blades start acting upon it. There is a great reduction in both the size and the mass. The cotton lumps are thrown over the grid bars due to high centrifugal force of rotating vertical Crighten opener. During this action, the tufts are greatly reduced in size. Simultaneously, the trash is liberated free and is forced through the gaps between the surrounding grid bars. Owing to reduction in their size and due to release of impurities, the cotton tufts become lighter in weight and are helped to rise by the tips of the striker blades which follow the spiral moving in the upward direction. The thrashing of cotton against the grid bars by the striker blades continues as the cotton rises up. The reduction in cotton mass and release of a large amount of impurities also continue.

An air current is provided to help and assist the cotton rise up, and it is created by a suction fan situated after the delivery duct. The suction fan induces a partial vacuum

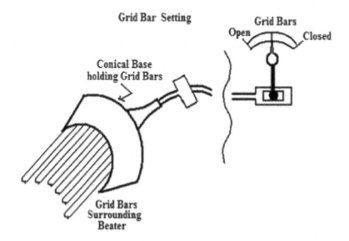

FIGURE 4.15 Angle of Grid Bar Angle[1,6]: This angle decides the percentage & proportion of lint and trash in the dropping.

and is surrounded by a screen. Owing to suction, the cotton is taken out of the machines through delivery duct. It is led on to the screen covering the fan. It is at the screen where cotton gets separated from the air which enters the fan and is suitably guided into filter bags. The partial vacuum created by the fan initiates and allows the required air to be let in through the feed pipe. However, the strength of the air current guiding the cotton within the beaters to move up has to be judiciously controlled. This is because it regulates the speed with which the cotton passes upwards through the machine. If the air current is too strong, the cotton simply passes through the beater very quickly and this allows comparatively less cleaning. Further, the air is also likely to be drawn through the grid bar spacing. Thus, the incoming air through the grid bars is very likely to oppose the out-going trash through the grid bars. The returning of the trash back into the beater chamber is likely to impair the cleaning action. Equally important is to see that the strength of air current is not too sluggish. This is because the cotton may not rise up through the beater at the desired speed. Eventually, it is quite likely to get over-beaten. This is far more dangerous as it would lead to more imperfections, especially the neps. It may also be noted that this over-beating makes the cotton more stringy. When exceptionally dirty cotton is processed (trash content more than 8–9%), it was customary to use two Crighten Openers in tandem (Fig. 4.16). In this case, by-passing flap at G is set accordingly. But sometimes, the airflow created by a fan positioned after the second Crighten is not sufficient to pull the cotton at the desired speed through two machines. Hence, an additional fast-moving cage (not shown in Fig. 4.16) with a suction fan is interposed in between the two Crighten Openers. When the two Crighten Openers are placed in this manner, it is important to provide a by-passing arrangement; so that either one of them or both (when processing finer grade cotton) can be omitted. The vertical beater is driven directly with the help of flat belt by individual motor mounted on the top of the framing. The production of the machine varies from 550 to 600 kg/hr.

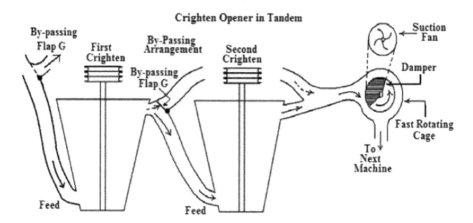

FIGURE 4.16 Crighten Openers in Tandem[1,6]: Being very harsh in action, when trash content was very high (above 8-10%), two such machines were employed, one after the other, in old days.

4.8.2.2   Factors influencing Crighten Opener Cleaning Efficiency[1,6]

1. The setting of the beater blades to the grid bars – If it is close, it tends to over-beat the cotton fibres and curls them. Therefore, along with the trash, the fibres are also forced through the grid bar spacing.
2. The state of cotton fed to the machine – If the cotton is not sufficiently loosened-up before being fed to this machine, the bigger and matted lumps find it difficult to rise-up. They, therefore get over-beaten.
3. Leakage through the machine casing surrounding the grid bars – If the air leaks through the outer casing, it tries to find its way through the grid bars into the beater chamber. Eventually, some of the lighter impurities try to re-combine with the cotton in the beater. As explained, this seriously impairs the cleaning efficiency of the machine

4.8.2.3   Periodic observations of droppings under the beater
The angle of the grid bars and their spacing should be such as to extract maximum trash with minimum lint loss. A periodic observation easily reveals whether the droppings are more whitish (lint loss) in nature.

4.8.2.4   Building-up of droppings under the beaters
An excessive piling-up of trashy matter under the beater not only impairs the beater cleaning action but also allows impurities to return back into beater chamber, thus recombining the lint and trash.

4.8.3   STEP CLEANER[1,6]

The Step Cleaner or Ultra Cleaner had been introduced much later and was supposed to be a very versatile cleaner. It beautifully combines the action of opposing striker blades against the grid bars. The opening and cleaning potential as compared with that in Porcupine or Crighten is much gentle and yet very effective. Hence, both heavy and lighter impurities such as motes, sand, leaves and stalks are removed from the cotton fibre mass without damaging the fibre. In fact, the cotton is never gripped and then beaten. Therefore, the blows of the striker blades are never harsh and yet, are more effective in removing trash. It is owing to this versatility that this machine can be used for both low grade as well as higher grade cotton with varying trash content. Like Crighten, Step Cleaner can be used in the early part of blow room line, and many a time, usually follows Bale Opener. Again owing to its typical versatile action, it is repeated in the blow room sequence and can be effectively used again in the later part of the blow room line. However, two Step Cleaners, in tandem, are never employed only because there is no suction used to draw the cotton through the Step Cleaner.

The machine requires an even and regular feeding and cotton is made to enter through a feed chute A (Fig. 4.17). There are, in all six beaters (B). As the material enters the region of the lowest beater, its striker blades immediately start acting on the cotton tufts. Each beater consists of a tubular body on which four rows of spikes or striker blades of elliptical cross-section are welded. The opening and cleaning take places when the cotton tufts are worked by the beater blades and are struck against the grid bars which partially surround the beater. All the six beaters are arranged with an inclination of 45^0 to the horizontal. As seen from the Fig. 4.17, all

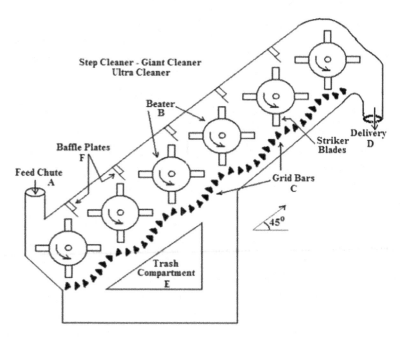

FIGURE 4.17 Step Cleaner[1,6]: Most versatile inclined beater used for both coarse and fine cotton with high and low trash content.

the beaters run in the same direction and this helps the cotton to rise up easily towards the direction of delivery chute (D).

Each beater flings the cotton on the sharp triangular grid bars which are adjustable. Both, the angle at which they face the beater and the spacing between them and the beater blades can be altered so that the proportion of lint and trash in the droppings under the beater (trash compartment E) can be adjusted by trial and error method. For low-grade cotton, the bars are set to face beater blades with their sharper edge; so also the distance between them and beater blades is narrowed. This permits maximum extraction of trash. On the other hand, when processing long staple cotton containing comparatively much less trash content, they can be set a little wider, closing the gap in between them. This avoids unwanted fibre loss.

Unlike Crighten, there is no induced air current to help the cotton rise-up. This, in fact, avoids any interference with cleaning process and lighter impurities are easily removed. However, there are baffle plates (F), interposed between each of the beaters. They baffle the circular air currents created by the beater rotation. These plates can be set along the beater cover so as to either lower or raise their position. The baffle plates divert these circular air currents in such a manner as to lead them to the next higher beater. This helps in passing the cotton to the succeeding higher beaters and prevents any undue lingering of cotton around a particular beater. It also keeps the uniform and continuous flow of outgoing cotton. The speed, with which cotton passes ahead and up, therefore can be regulated by shifting the position of the respective baffle plates. Thus, when the plates are pushed closer to lower beater, the cotton moves-up more quickly.

In this machine, there is a combination of mechanical spike – grid bar action. Along with this, the centrifugal force of the beater blades throws the cotton tufts on to grid bars and effectively removes the impurities, which after passing through the grid bar-gaps fall down by gravity, in the trash compartment situated below. A progressive increase in the beater speed, from lowest to the highest, results in greater extraction of trash. Most commonly, speed of around 550 r.p.m. is used. At this speed, the proportion of lint and trash in the droppings is found to be most economical. This gives the best cleaning efficiency. However, when the speeds are increased, there is a sizable increase of proportion of lint in the droppings with only a little increase in the trash content. It was always argued that this beater cannot match its cleaning performance with other beaters. In fact, in old times, it was labelled as 'half cleaning point'. The research work in this case has proved that the cleaning performance of Step Cleaner is not less than any other normal beater as expected cleaning efficiency of this machine is around 30–35%, a value not much less than Crighten Opener. It may also be noted that when this beater is placed in early blow room line, its performance is far superior to what it is when placed in the later stages of blow room.

4.8.4 Axi-Flow[1,6]

Unlike Step Cleaner, the passage through this machine is very different. In fact, it is different from any other machine. In Step Cleaner, the cotton moves through the

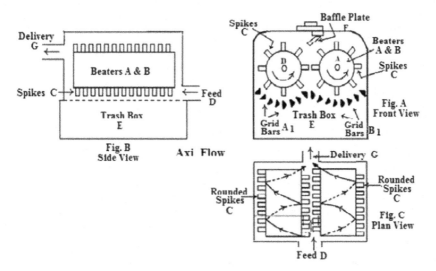

FIGURE 4.18   Axi Flow[1,6]: It is a versatile opener used beneficially for cotton with both high and low trash content.

machine along the direction of rotation of the beaters. In Axi-Flow, it moves across, almost perpendicular to the plane of direction of the rotation of the beaters. The mechanical spike-grid bar action and the centrifugal force, however, are combined in the same manner to remove the trash. The machine has two hollow cylindrical beaters A and B (Fig. 4.18) which are placed horizontally and very close to each other. The blades are welded on both the beaters and they cover the whole circumferential area along the full length of the machine. The blades are round in cross-section and induce buffeting and agitating action on the incoming cotton.

Both the beaters rotate in the same direction and each has its own system of grid bars A_1 and B_1 beneath them. However, as can be seen, at the point of their intersection (a common junction point between them), their directions oppose each other. Like Step Cleaner, some pre-opening is required for Axi-Flow and hence, after two or three initial machines, it may be beneficially placed in the blow room sequence.

A condenser cage is placed subsequent to this machine with a suction fan, which creates partial vacuum and thus induces air currents to pass through the machine. This regulates the speed of cotton passing through the beater section. The cotton enters the beater chamber through inlet or feed (D) and comes immediately under the powerful action of both the spiked cylinders rotating at nearly 400–450 rpm. Owing to very strong sweeping action of the blades, there is reduction in the size of cotton mass. The blades throw the cotton on to the grid bars and impurities liberated free are thus released through them.

The bar angle can be adjusted so as to arrive at correct proportion of lint and trash in the droppings. The best setting is always achieved when there is minimum lint loss with maximum trash extraction. The trash which drops down is collected in the trash box (E).

As the cotton has to move in a perpendicular plane to that of the rotational plane of beaters, the feed and the delivery positions are displaced (side view in the figure). This does not allow the cotton to directly travel through the beater and get influenced by the suction of the fan at the delivery end (G).

It ensures that cotton is led to the delivery duct only after being thoroughly treated by beater blades. Thus, a very good opening followed by an effective cleaning is obtained. An adjustable baffle plate (F), positioned at the top, deflects back the cotton, which is flung upwards by beater A, into an area above the point of common intersection of the two beaters. This prohibits again the same cotton to be simply carried by the suction of the fan.

The air stream passing through the machine is conveniently kept above the action zone of the beaters. This avoids any interference with cleaning carried out by the beaters. The trash box placed under the beaters is provided with an inspection window and the richness of trash or lint in the droppings can be viewed during normal running of the machine. Within the machine, the cotton doesn't have to move against the gravity much; however, the rotation of the beaters does not help in any way, the cotton to move ahead. Therefore, unlike Step Cleaner, where there is no air current helping the cotton to move up, in Axi Flow, much stronger air current is required to control over-treating of cotton within the beater chamber. However, in both, Step Cleaner and Axi Flow, the cotton is acted upon in its free flow, and worked by the beaters. Hence, the cleaning action of both is almost similar.

The production of this machine is approximately 550 kg/hour at normal level of 20–25% cleaning efficiency. The Indian version of Axi Flow is 'Spiro Cleaner', which is also very popular in mills owing to its versatility. However, when the immature fibre proportion is high, there is higher nep generation and higher loss of fibres in the dropping. The great advantage of this machine is that it requires very less space and the power consumption and maintenance cost are much lower. A point to note here, is that unlike porcupine opener, the cotton has to travel across the machine (not helped by the beater blades). Therefore, when too much matted mass of cotton is fed to the machine, it may get jammed within the beater chamber. The beater, therefore, has to be never placed immediately after bale breaker but always a little later.

4.8.5 MONO-CYLINDER CLEANER

Increasing use of mechanical pickers during harvesting and bad conditions of ginning machines for seed separation results in considerable high trash content in the cotton that arrives in the mills. With lower grade cotton, higher trash content may not pose serious problems; because the fibres are shorter, coarser and comparatively stronger. During processing they can sustain intensive blow room action. However, with fine and superfine varieties of cottons, higher trash content necessitates gentle treatment and yet effective trash removal. Also with them, the blow room sequence of machines needs to be much shortened. The use of Mono-Cylinder Cleaner (Fig. 4.19) provides the right choice. It gives gentle treatment and yet satisfactory cleaning.

The gentle opening with effective cleaning is provided by the machine because, the stock is struck by the mono-cylinder striker blades when the material is not

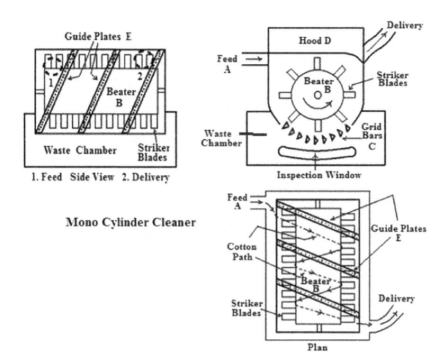

FIGURE 4.19 Mono Cylinder Cleaner[4,6]: Like Axi-Flow, the material moves perpendicular to the beater rotation. It is effective in removing trash, not harshly, but comparatively gently.

gripped. In fact the striker blades act on the cotton tufts when they are in the free flow. Thus, the impurities are released when the blades open-out the tufts and throw them on the surrounding grid bars. Further, the impurities are not crushed.[I] The machine also avoids overbeating the cotton and which again reduces nep formation.

Mono Cylinder Cleaner is most conveniently employed after initial opening of bale cotton, preferably after a few machines. This helps in getting the best contribution from this machine. The cotton enters (Fig. 4.19) the machine at (A), at right angles to Beater (B). The cylinder itself is hollow circular tube. There are eight striker blades around the circumference while along its length there are many striker blades. These blades immediately take hold of entering cotton and start working on it against the grid bar (C). This results in both very good opening and release of considerable amount of trash which is ejected through grid bar openings. To assist the movement of cotton through the beater, a suction fan is situated along the delivery chute and close to the next machine in the sequence. It induces air currents through and across the cylinder.

The whole cylinder is covered with a hood over it. This hood has guide plates which are fitted on its inside portion. They are positioned over the cylinder. This helps in making the cotton tufts travel in spiral form from the entry side to the delivery side

---

[I] Impurities when crushed spread over and become more difficult to be removed.

of the beater. Thus, when the tufts are flung on the hood owing to the cylinder rotation, these guide plates do not allow the cotton to come directly under the influence of air currents which otherwise would have carried the stock, directly into delivery chute. The guide plates on the hood thus help in sending the cotton back on the cylinder below. In this way, the cotton, while moving spirally gets flung again and again on the grid bars below, and in the process gets cleaned. The action of the blades on cotton is never harsh and still the beater cleaning efficiency is quite good.

The delivery chute is arranged to be tangential to the cylinder circumference so that the cotton tufts can emerge along their natural exit path, i.e. along the direction of cylinder rotation. This helps in avoiding any eddy currents. The cylinder is fitted with ball bearings at both ends to ensure smooth running at high speeds.

The grid bars underneath are adjustable for their opening and this facilitates controlling the lint and trash proportion in the droppings. The hood itself can be opened for any maintenance operation. In the normal running condition, there is locking arrangement provided for the hood for safety. The doors are provided to waste chamber for the removal of the trash. The whole trash chamber is illuminated from inside and through a glass window, the observations can be made for seeing the nature of trash. The cylinder is driven by V-belt, using 3 H.P. motor. The machine with its single cylinder occupies a very less space and the higher production rates are possible to suit the general production requirements of other machines in the sequence.

Salient Features of Mono Cylinder Cleaner:

1. High cleaning efficiency with gentle handling of cotton fibres
2. High production rates up to 800 kg/hour
3. Higher cylinder speed – around 700 rpm
4. Moderate air requirement to build the necessary flow for the effective transportation of the material through the machine is 0.7–1 cub. meter of air per second at 10 mm water column pressure
5. Adjustable grid bars – Change in angle possible from 20° to 30°, corresponding change in their spacing from 20 to 26 mm

4.8.6 ERM CLEANER

The machine is introduced by m/s Laxmi-Rieter and is very simple in construction. The opening capacity of this machine is very high owing to the use of saw-tooth wires or toothed discs on the beater (A). Different types of beaters are available to permit the quick adaptation for variety of raw materials. This permits greater flexibility in its functioning. The beater section is newly designed and consists of spring loaded feed roller system (B) to ensure uniform and controlled feeding (Fig. 4.20). The spring loading facilitates sufficient hold on the material issued and at the same time, avoids any excessive nipping. It prevents the snatching and yet enables the feeding of cotton in sufficiently loose form. Thus, a higher degree of opening and cleaning is possible by still maintaining gentle handling of cotton.

The material is received from the previous machine through a chute (C) by using a suitable suction device. It is then separated from the influence of air currents and

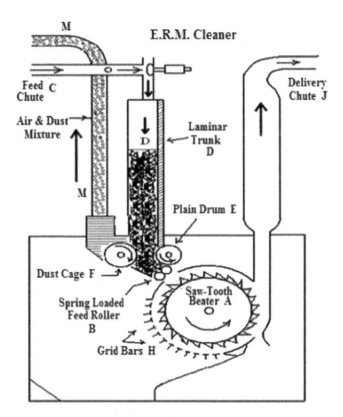

FIGURE 4.20   ERM Cleaner[4,5]: A saw tooth cleaner leads to very effective opening of the cotton tufts.

dropped into a laminar trunk (D). The head of the cotton level in the trunk is controlled electronically. A large glass window in the front of the trunk is provided for the observations. The material at the bottom is guided over the plain drum (E) and perforated dust cage (F) to filter and extract the fine dust from the loose cotton piled-up within the trunk. The air through the cages is finally let out through another vertical chute (M) connected to the cage.

The grid bar section (H) is equipped with an adjustable sheet metal guide interspaced between the knife bars to control the gaps and also to provide for the passage of material. There are in all about 8–10 robust knives with metal sheet guide. The grid is given a fine finishing surface so that the fibres do not adhere to it. The cotton being broken down to very fine tufts is finally taken through the delivery chute (J). Here again, a suction device suitably placed subsequent to the beater helps. The suction air is filtered, whereas the cotton is delivered to the next machine.

It is claimed by the manufacturers that the smallest possible size of the tuft that comes through the machine weighs not more than 7–10 mg. For this, however, the state of material fed to this machine has to be carefully controlled. The cotton before entering the machine must be in sufficiently loose form, and this necessitates its placement in the later part of blow room sequence. If placed much earlier, not only opening and cleaning get affected but also it endangers the life of saw teeth of the beater. In fact, this is a common criticism levelled against any cleaner equipped with saw teeth. Therefore, the position of such beaters has to be carefully selected so as not to harm the tiny saw teeth. This is because; the tiny saw teeth can very effectively continue their opening action when the cotton fed is in loose form.

The beater with 15¾ inch diameter runs at 1000 r.p.m. The production rates are usually controlled up to 500 kg/hour. An erection platform is provided to allow the beater to be positioned or taken out quickly and easily. The machine has fewer parts, lower power consumption (beater – 5.5 H.P.; feed roller and gearing – 0.25 H.P.; and fan – 4 H.P.), a smaller filter installation unit and requires a moderate space of 38 × 59 (inch) with a height of 13 ft.

4.8.7 SRRL OPENER

The machine is basically intended for giving very intensive opening. With this opening, it is possible to easily extract the trash, especially the one which contains higher amount of vegetable matter. Within the machine (Fig. 4.21), there is great rolling momentum given to the large mass of cotton. All this takes place in a hopper or bin (A). However, this rolling, tossing or churning given to the cotton is conducive to high generation of neps, especially when treating a mixing with higher immature fibre proportion. As some of the types of Indian cotton contain substantial amount of short and immature fibres, the nep generation is quite high when this machine is used. Even then, the action of the opener is never harmful to cause any damage to the fibres, and hence, for better varieties containing higher proportion of mature fibres, this machine could be very useful owing to its high potential for very effective opening of the tufts.

The cotton may be fed to the machine by either blending hopper or conveyor belt (B). Usually a metal detector is fitted very close to the conveyor belt to prevent any in-going metal part that would damage the saw-tooth rotating cylinders (E).

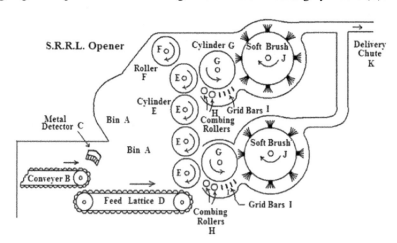

FIGURE 4.21 SRRL (Southern Regional Research Laboratories) Opener[1,6]: A very good opener but comparatively mild cleaner.

The lattice (D) is the feed lattice with which the cotton is brought into bin (A). The saw tooth cylinders (E) pick up small quantity of cotton at a time, through their tiny teeth and carry them upwards. The speeds of cylinders (E) are progressively increased in the upward direction.

This ensures an even distribution of cotton picked up by the lowest cylinder on all higher cylinders (E). The bottom cylinder rotates at 340 r.p.m. while the one at the top runs at 400 r.p.m. The lattice (D), cylinder (E) and uppermost roller (F) – all together create a rotary motion to the mass of cotton in the bin (A), New mass is constantly picked up from the lattice D by the lowest cylinder and is brought into churning action.

The cylinders E, with their small saw teeth, pick up equally small and tiny mass of cotton and this results into very intensive opening. There are two other cylinders (G) which are also covered with fine saw teeth and are closely spaced with cylinders (E). The cylinders G, thus pick up the tiny mass of cotton from rotating cylinders (E) and work it over respective combing rollers (H) and grid bars (I). Some cleaning is expected to take place here. The trash thus released falls down through the grid bar openings and is collected into the trash box placed below.

There are two revolving brushes. They have soft bristles just touching the respective cylinder (G). The brush bristles are able to pick up the highly opened out and very small cotton tufts from the surface of each of the cylinders (G). A radial air stream created by a fan placed subsequent to the machine generates strong air currents which flow past the brush bristles and thus sweep the cotton away. Finally, the cotton is led out of the machine through the delivery chute (K).

The most important maintenance work involved here is in keeping the saw tooth wire points without any damage. Their effective action ensures high level of opening. This makes it necessary that the condition of cotton entering the beater is sufficiently loose. There are instances where the wrong placement of this beater has led to frequent damages of saw-tooth wires. In this context, a much earlier positioning of this machine in blow room sequence would prove to be detrimental, especially when high density bales are processed. The production of this machine is around 650–700 kg/hour. The cylinders and lattice are driven by 9 kW motors. Owing to more focus given in the machine for opening, the cleaning efficiency is expected to be lower.

4.8.8 RIETER'S STRIKER CLEANER

In this machine, the advantage is taken of gravity to lead the cotton through the machine. The three beaters are placed in a zigzag manner (Fig. 4.22). At the top, there is a condenser cage B with a suction fan that induces very strong air currents and draws the cotton from the previous machine through an inlet pipe (A) onto the cage.

As against Step Cleaner, where the cotton has to move against the gravity, the cotton through this machine falls freely down the beater chamber and is worked over by the beaters during its downward journey. The treatment that the cotton receives, therefore, is much milder. Thus, the machine is suitable for both short and long staple cottons. However, as compared to Step Cleaner, its cleaning efficiency is comparatively lower. The cotton through inlet pipe (A) enters the machine and is loosely collected over the cage surface. The cage rotates at a high speed of

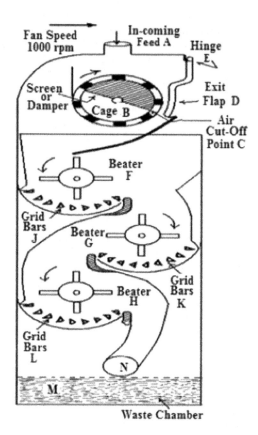

FIGURE 4.22 Striker Cleaner[4,6]: Another useful cleaning machine using gravity for transporting cotton through it.

35–40 r.p.m. and is provided with a screen or damper which is placed inside it. This damper (shaded area in the figure) restricts the suction to an area facing the inlet pipe (A). Thus, beyond this area, the suction is not allowed to have any influence over the cotton deposited on the cage. The small perforation of the cage not only helps in drawing air but also provides a passage for the tiny trash particles and fine dust, both of which are effectively sucked in.

The rotation of the cage helps the cotton to glide down. During this movement, the cotton pushes the exit flap (D) around its hinge (E). It then slides over a curved surface and falls onto the fast rotating beater (F). Suitable guide plates are provided to guide the cotton within the machine to lead it from beater F to beaters G and H. The beaters have strong striker blades with rounded bars of 9–10 inches length. The grid bars underneath each beater are quite sharp and work as the cleaning elements.

It may be noted here that the cotton is not gripped during its journey through the machine. Therefore, the action of the beater blades does not harm the fibres even at high rotating speeds. Even then, each of the striker blades is very effective in picking up the cotton tufts and striking them against the respective grid bar systems (J, K and L). This results in very useful cleaning. All the beaters run at 350–400 r.p.m.

The trash, thus, released falls down through grid bars systems into trash chamber (M). The grid bars are adjustable for their opening and the angle with which they face the beater blades. This provides some flexibility in the cleaning operation. A funnel is provided at (N) with a delivery piece which carries the cotton side-ways and leads it into delivery chute. The normal production rates make it possible to work this machine at 750–800 kg/hour.

4.8.9 SHIRLEY OPENER

It is another saw tooth type of beater which gives a very intensive opening action to the mass of cotton. The peculiarity of saw tooth beaters is that the tiny teeth are able to bring about much-intensified opening of cotton tufts. This opener is normally placed in the later part of the blow room sequence. The only care that needs to be taken for its efficient performance is the controlled feeding. This is very much essential to avoid any over-burdening on tiny saw teeth at any time. With a controlled feeding, however, the beater can be very effective in opening the cotton tufts to smallest possible size.

The cotton is fed (Fig. 4.23) on the lattice (A), which carries it forward with the help of assisting rollers (B). The pair of fluted feed rollers system (D) is given additional weighting on either side and presents the cotton in more compact sheet form. There is also a pedal (E) and feed roller system which further regulates the amount of cotton entering the beater zone. The pedals are fulcrumed around the bar (F). They carry long tails and corresponding dead weights. These weights always press the respective nose of each of the pedals on the pedal roller.

With this kind of feed control, the cotton enters beater zone. The cylinder beater (J), having a diameter of 15.625 inches, is covered with tiny saw tooth wires and runs at 1560 r.p.m. The saw tooth wire-point density is approximately 24 teeth per sq. inch. The fibres carried by the sharp and tiny saw teeth are struck on the knife edge (K). A high degree of opening is achieved leading to the release of large amount of trash.

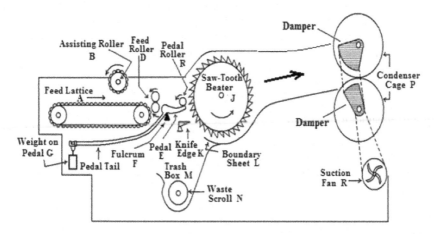

FIGURE 4.23 Platt's Shirley Opener[1,6]: A fine saw tooth opener similar in design to Shirley Analyser and uses Aero-dynamic principle for separation of lint & trash.

This trash is collected in the trash box placed underneath. A slow rotating waste scroll roller (N) periodically pushes this gathered trash at the side of the machine from where it is taken out.

A suitably placed boundary sheet (L) is instrumental in separating the lint and trash very effectively. The cylinder beater J, owing to its high rotational speed, creates concentric air stream. The trash being heavier than lint has less buoyancy and follows the outer rotating air-layers around the saw tooth cylinder. The boundary sheet when correctly positioned, very efficiently separates this trash to fall away and below into trash box.

The setting of the boundary sheet (L) is important in this respect. When set closer to beater J, more trash falls down and it also takes more proportion of lint along with it. As against this, when it is set farther away from beater J, a sizable proportion of lint is saved. However, there is comparatively less proportion of trash in the droppings. The setting of boundary sheet thus controls both the cleaning efficiency of the machine and lint loss.

Therefore, the setting of boundary sheet has to be made judiciously taking into consideration both, the cleaning efficiency and the proportion of lint in the trash box. The cotton material finally gets deposited on the fast revolving cages (60 r.p.m.) and is delivered to the next machine in the sequence. The normal production rate is 60 kg/hour. Though the action of the beater is powerful, there is no fibre damage caused. Therefore, the machine can also be used for better grade cotton varieties.

4.8.10 AIR-STREAM CLEANER

This machine also uses aerodynamic principle to separate lint and trash. As mentioned earlier, the lint being lighter has much higher buoyancy and hence very low inertia. This is very beneficially used to separate lint and trash in this machine.

A high velocity stream is used to carry the mixture of lint and trash, both of which are very effectively separated. This separation is achieved by using a modified version of Kirschner type of beater (Fig. 4.24). The only difference here is, unlike the fine needles of Kirschner, very fine saw tooth wire is clothed on the staves of cylinder in this machine. The velocity of the stream of air carrying the mixture of lint and trash is further increased with the help of a booster fan (I). The lint and trash thus travel at a very high velocity. It is at this point when the path of the mixture of lint and trash is suddenly changed. Owing to high inertia and low buoyancy, the trash does not follow this sudden change. This is advantageously used to separate it from lint.

Kirschner type of beater (A) is situated prior to actual device. The saw tooth points on this beater are very effective in giving thorough opening of the stock fed. The knife (D) helps in further improving this opening when some trash is liberated free. The boundary or separating sheet (C) is adjustable and can be set with the beater.

Thus, the closer setting allows more trash to get diverted at the point D to go down into the trash box. As stated earlier, this allows some percentage of lint to also accompany the trash. However, the cotton-lint at separating sheet, to a certain

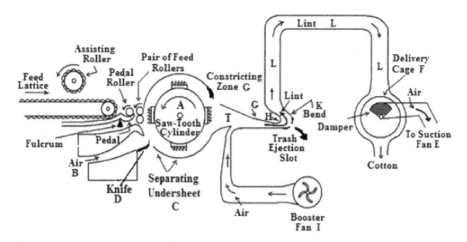

FIGURE 4.24 Platt's Air-Stream Cleaner[1,6]: It precedes saw tooth opener & carries out very fine opening of the cotton tufts. The aero-dynamic separation which follows gently separates the trash from the lint.

extent, is supported by a mild air current which enters at (B). The mild current is generated owing to the rotation of the beater itself. This mild air current follows a path around the beater and helps in reducing the lint loss at (D). In any case, owing to the saw tooth action of the beater, the cotton is taken around the beater and is in much opened condition.

The cotton in this state is delivered into a delivery chute (T). It may be noted here that, even when the beater with the help of knife and separating sheet is able to carry out certain cleaning, the cotton entering the delivery chute still has some trash leftover. Thus, the material in the delivery chute is a mixture of lint and trash in very loose and opened condition. A strong and powerful air current which is generated by a booster fan (I) is introduced into the delivery chute and enters the chute just at the beginning of the delivery chute and after the action of beater (A). This strong air current carries the opened and loosened-out mixture of lint and trash through the delivery chute at a great speed. The chute is narrowed down further along its passage (G). This narrowing of the chute continues till H and due to this; the velocity of the mixture of lint and trash through the tube enormously increases. The velocity around the narrowest part in the chute (around H) is almost 55–60 m/s. At this point, the chute path is suddenly changed through almost 120^0 from its original direction.

The aero-dynamic separation is based on the difference in buoyancies of lint and trash. The trash being heavier has low buoyancy and higher inertia. Hence, the trash particles cannot follow such a sharp turn at bend K. Owing to inertia, they are deviated only a little; whereas, the lint being much lighter easily takes the sharp bend and follows the chute further. An ejection slot is provided anticipating the track of trash around the bend and is adjustable. By setting the opening in the slot, a free-way can be provided for the trash to pass through this opening. Thus, the trash is ejected through the slot and gets removed.

The cotton fibres follow the main air stream and travel along pipe (L). They are finally led further onto a delivery cage (F) which is provided with a powerful suction fan (E). The air conveniently passes through the cage perforations, while the fibres merely get deposited over cage surface. Even here, very fine dust that accompanies the air gets removed through the perforations.

The axial fan E while creating a powerful suction also helps in reducing the pressure of rushing air into the trunk L. This helps in streamlining the airspeed between regions G and H. Further, when the ejection slot has very small opening, the force of air in this region is very high. This results in only small amount of trash being extracted. The trash thus extracted mainly consists of broken-crushed leaves, stalk, sand and only a little proportion of spinnable fibres. As the ejection slot opening is widened, the speed of air is slightly reduced and even the lighter impurities get extracted. However, this is associated with a small and yet sizable proportion of lint.

With higher airspeed resulting from the action of the booster fan (I), the impurities embedded within cotton tufts also get partially released and cleaning action of the machine is further improved. The setting at the knife edge (D) and circumferential and radial setting of under-sheet (C) also control the amount of waste extracted under the beater.

The cleaning efficiency of Air-Stream Cleaner is around 30–35%. It mainly depends upon the opening of cotton prior to entering the machine and also opening carried-out by the beater itself. Owing to this, the machine is positioned much later in the blow room sequence. The air exhaust is around 98 cub.m (3500 cub. feet) per minute and it avoids any adverse effect of development of backpressure in trunk L. This makes it necessary that the exhaust air through the delivery cage (F) is led into a good setting chamber and not into the dust bags. This is because the dust bags are likely to get choked-up, from time to time, with the deposition of the fibres and the dust. The production of the machine with Kirschner-type of beater is around 400–500 kg/hour which is quite sufficient to feed two scutchers in the final part of blowroom line.

4.9   BLENDING IN BLOW ROOM

The conventional cotton blow room line with a single processing (no manhandling between bale to lap conversion) is equipped with blenders which resemble hopper feeders in construction and yet, serve an altogether different function.

The pins of inclined lattice of blenders, though much finer as compared to that of Bale Breaker, are strong enough to break open the matted lumps of cotton from bales.

In conventional blow room sequence, there used to be 4–5 blenders arranged (Fig. 4.25). The cotton bales were laid around each of these blenders near the feed region. The operatives were directed to take thin flat slices in the form of slabs from each of the bales. They were laid on the creeper lattices. One of the blenders was specifically used for feeding soft waste produced in spinning. The soft waste in spinning mainly consists of lap and sliver pieces, sliver web at various places and bonda waste, the last being the softest and richest possible waste collected in the

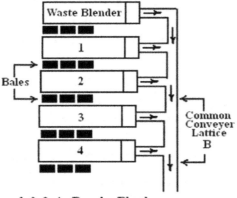

1, 2, 3, 4 – Regular Blender

FIGURE 4.25 Automatic Blenders[1,6]: Basic tuft-opening machine – opened-out cotton tufts from individual blender are laid on a common lattice which facilitates easy mixing of material.

whole of spinning department. It is the pneumafil suction waste at ring frame when the broken ends are sucked away. However, the proportion of this soft waste added is required to be restricted to less than 2% to ensure satisfactory working in the subsequent processes. After processing the bale cotton through blenders, it is led on to a common conveyor belt (B) running across all the blenders and thus collecting the material from each of them. The belt, after collecting the material lets it into a duct which pneumatically carries it to the subsequent machine. With trends for shorter processing, the blenders do a thorough job of mixing different lots of bales.

If there are 'D' bales laid for a particular mixing and 'S' is the weight of the smallest slab that the tenter is asked to pick up from each bale, then the total quantity of material put into the hopper of the blenders during one round of the tenter is –

$$= D \times S$$

As can be seen, this can never be greater than the combined capacity of the hoppers of all the blenders taken together.

If the hopper capacity of each blender is 'H' and the total number of blenders used are – 'N', then the combined capacity of all the blenders used = H × N

In a simple equation form the above can be put as follows:

Weight of each Slab × Total Bales ≯ Each Hopper capacity
        S   D   H
× Total Blenders
        N
OR simply as, S × D ≯ H × N

The right-hand side of the above expression, for a given installation is constant.

For example, if N = 5 blenders; H = 20 kg; then the value of the right-hand side of the above expression = 100 kg Thus,

$$S \times D \ngtr 100 \text{ kg}$$

With the above equation, it can now be realized that the product of (S x D) should not be greater, in any case than 100 kg. This is because it represents the total hopper capacity of the blenders used. The real importance of taking small slabs at a time by the workers can now be understood. For example, if the mixing is constituted by taking at a time (say) 10 bales (D = 10), then the weight of each slab can be up to 10 kg. However, if the number of bales increases (say 20), the slab weight cannot be more than 5 kg. In any case, taking smaller slabs from each bale at a time is always better because then all such slabs, when arriving in the hopper region, will have good chance to mix intimately with each other.

As explained earlier, with a large number of bales in a mixing, it would ensure that the variations in fibre properties are spread more uniformly within a mixing lot. Thus, the more the number of bales, better is the uniformity of fibre mixing. Therefore, when the value of 'D' is higher, the proportion of cotton taken from each bale for every round (slab-weight), should be restricted to only a small value. The operatives often have a tendency to take bigger slabs to flood the creeper lattice with large mass of cotton. This gives them some freedom to rest in between their working rounds. However, larger slabs from each bale do not give adequate opportunity for cotton from each of the slabs to be properly represented in the mixing-hopper. Agile supervision is, therefore, necessary to see that the operative picks up only a small quantity from each bale. Using more number of blenders has also an advantage, as the load on each blender is then reduced and the opening action in each is improved.

4.9.1 AUTO MIXER

With the demand for better quality and evenness of the yarn, the necessity of uniformity and homogeneity of a mixing has increased. Like blenders, this machine – auto mixer (Fig. 4.26) – is very effective in giving a thorough blending of different lots of cotton. It works on the principle of 'stack mixing'. The advantage of auto mixer is that it is very useful when one wants to shorten the blow room sequence with more effective opening and cleaning. The normal position of this machine is after the bale breaker.

The condenser cage (C), with its suction fan, collects the cotton from bale breaker. It then delivers the material onto a conveyor belt (B) of the distributor. The belt and the distributor are mounted on a trolley which is made to run on a suitable railing in either direction. With the trolley moving to and fro, the belt also runs along the direction of the trolley. Thus, the material on the belt is suitable dropped down into a mixing compartment (D) placed underneath. With about 30–40 layers (which can be pre-set), the whole stack is build-up in the mixing compartment D.

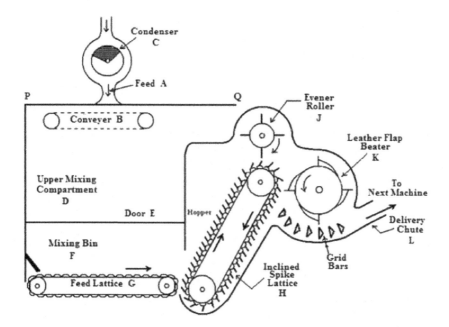

FIGURE 4.26 Auto Mixer Principle[6]: The machine is designed on stack mixing principle where the stacks of different mixings are laid in a compartment and the sections are taken vertically to represent different mixing layers.

After the desired number of traverses, the distributor stops and the door (E) opens to allow the mass laid within the compartment to fall into a mixing bin (F). Immediately, the bottom feed lattice starts moving in the direction (shown by the arrow) ahead towards inclined lattice (H) and brings the cotton closer to it. The cotton in the layer form in mixing bin F is worked over transversely across the vertical height of the stack by inclined spike lattice.

This ensures that the material picked up by the inclined spikes represents cotton from different layers (and possibly different lots) as gathered in bin F. The subsequent working is similar to that of Hopper Opener and the Evener Roller (J) and the Leather Flap Beater (K) function exactly the same way to finally deliver the cotton through the delivery chute (L). The arrangement is made so that no cotton falls from the compartment D to F till the cotton in the latter gets completely exhausted. During the time, therefore, when the earlier cotton in mixing bin F is being worked, the compartment D is closed.

During the time, when the material in the compartment F is being worked, the feed to the compartment is, however, continued with the next batch of lots. For this, the feed through condenser cage is started again. The fresh stack mixing is prepared in the closed compartment D and is kept ready for working the next lot of stack mixing. Even when the mixing is to be changed, the corresponding bales are kept around Bale Breaker and the material from it can be brought through condenser and laid into layer form in compartment D. Thus, the door E is kept closed till all the previous material in the mixing bin F gets completely exhausted. The moment it does, the door E immediately opens and lets the new mixing lot to fall down into mixing bin F. This avoids considerable loss of time during each mixing change.

4.10   BLADED BEATERS

By the time the cotton is processed through several opening and cleaning machines in a blow room sequence, a substantial amount of trash gets removed. However, it still contains some residual trash depending upon the cleaning efficiency of the earlier machines. Similarly, at the fag end of blow room, it is also the time to think of building a well-textured sheet of fleece of cotton in the form of a lap.[II]

The bladed beaters meet both these purposes. On one hand, they continue the cleaning carried by the previous machines and, on the other hand, they deliver the material in a sheet form.

Thus, their position in the blow room sequence of machinery gets automatically selected as the beaters in the ending part of blow room. The bladed beaters have a unique action on cotton. Unlike any other beater discussed so far, the blades of this type of beater extend across the full width of the machine, and hence these blades simultaneously act on the full width of a cotton sheet. Needless to mention here that the beater, therefore, demands feed in a loose sheet form. The cleaning ability of these beaters depends upon their beating power with which they strike the cotton sheet.

The final part of the blow room line is called 'Scutcher'. It is mainly responsible to form a final lap. The bladed beaters are part of this scutcher. To make a uniform lap, apart from the cleaning that bladed beater can provide for, it is equally important to open the cotton tufts to a very small size. An outstanding feature of bladed beaters is that each blade strikes very powerfully on the sheet of cotton presented to them and reduces the tuft size.

A feed system A (Figs. 4.27 & 4.28) carries the job of issuing uniform cotton sheet. This sheet is quite loose and not consolidated. As the beater blades extend the full width of the machine, they are able to strike this cotton sheet simultaneously and fully across its width. The feeding system has a firm hold on the cotton sheet when issued to the beater. This results into very effective opening of cotton tufts. A significant amount of trash is released when the beater blades strike the cotton onto the grid bars positioned below. The bladed beaters are unique in extracting the motes with the fibres attached to them and in this respect, their action is unique.

The grid bar section (C) is mounted in a curved bracket with appropriate slots for each of the triangular grid bars which are adjustable for their angle with which their sharp edges face the action of the blades. Along with the angle, the opening between

---

[II] In conventional blow room, the final product was a rolled cotton lap sheet. In modern installations, the cotton is directly fed to the next machine in the loose form – Chute feeding System.

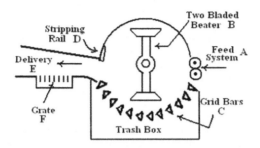

FIGURE 4.27   Two Bladed Beater[1,6]: It gives very heavy & forceful beating to the incoming cotton and throws it on to the grid bars to materialize trash extraction.

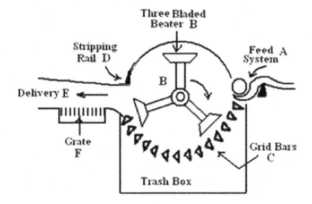

FIGURE 4.28   Three Bladed Beater[1,6]: It is another powerful beater giving forceful blows on cotton tufts. Many impurities get extracted through grid bars.

the grid bar is also adjusted simultaneously. Both the angle of the sharp edge and the opening between each grid bar decide the amount of trash thus released. The trash, after passing through the grid bars, is suitably collected into the trash box placed below.

Sometimes, a controlled airflow is allowed to pass through a later portion of the grid bars from underside in an upward direction. This helps in smoothly allowing the cotton to pass through the beater chamber. Thus, the cotton is not allowed to unduly linger. This airflow, however, has to be controlled very carefully. This is to see that this controlled airflow does not carry back the trash which is expected to fall down through the grid bars. Any negligence in this is likely to impair cleaning carried-out at the beater. Some machinery manufacturers admit this air at a point beyond the two-third distance from the first bar. In this case, the trash box needs to be divided into two compartments, the initial two-third compartment being air-tight, so that normal beater cleaning is not affected. The subsequent one-third compartment is allowed to be equipped with an inlet valve to control the strength of air current entering the beater chamber.

There are three important settings – (1) Striking Distance of feed system (2) Setting of grid bars and (3) Stripping rail (D) – all are required to be set with respect to beater.

4.10.1 STRIKING DISTANCE

In the feed system, the distance between nipping point of pair of feed rollers or pedal & feed rollers and striking point of the beater blades (Figs. 4.29 & 4.30) has to be slightly greater than staple length of the cotton processed.

The striking distance, as a rule, has to be greater than the staple length. In short, when one end of the fibre fleece, is gripped by the feed system, its other end should not be simultaneously struck by the beater blades. This avoids any possible fibre damage. It can also be seen (Figs. 4.29 & 4.30) that in the case of the pair of feed rollers, the minimum distance is decided by half the diameter of the rollers, whereas in the case of pedal & feed roller, the distance is directly realized from the nose of the pedal.

Eventually, as can be seen from these figures, the actual setting carried out, is BC in the cases of the pair of feed rollers and DE with pedal & feed roller. It can also be seen that DE can be made much shorter as compared to AC. Hence, the pair of feed roller system is only used for longer staple cottons; whereas the pedal & pedal roller system is more appropriate and hence suitable for short staple cottons. Equally important is to see that the striking point (line C or E) is not set too wide; otherwise the blades will merely snatch bigger tufts of cotton from the feed system, and this will seriously affect both the opening and cleaning action of the beater.

4.10.2 GRID BARS

The distance of the grid bars from the beater blades regulates the intensity of striking action, and hence controls the level of extraction of the impurities. In some cases, it is possible to change the angle of the grid bars. Thus, the grid bars can be made to face the beater blades more tangentially. This increases their effectiveness in extracting more trash. However, as mentioned earlier, this also allows more lint in the droppings.

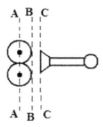
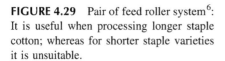

FIGURE 4.29 Pair of feed roller system[6]: It is useful when processing longer staple cotton; whereas for shorter staple varieties it is unsuitable.

FIGURE 4.30 Pedal & Feed Roller[6]: For shorter staple, the system is used for ensuring better gripping distance.

4.10.3 STRIPPING RAIL

The stripping rail D (Figs. 4.27 & 4.28) is situated very close to the upper casing of the beater on its delivery side. Thus, its position is very strategic. The setting of the stripping rail controls the deflection of air currents away from the beater and does not permit the cotton to go round-and-round the beater. It, thus, avoids any possible repeated beating of cotton within the beater.

Finally, as the cotton leaves the beater through passage (E), it passes over the grate (F), where the left-over and loosened out impurities may fall down through the small openings of the grate. Normally a perforated, rotating cage is placed at the end of passage E. A suction fan is suitably mounted to draw air through the perforations of the cage. Along with air, very fine impurities and dust particles enters through these perforations, whereas the cotton tufts are gathered over cage surface and once again, are transformed into a cotton sheet.

4.11 CAGES

The cages are thin perforated sheets forming the hollow drums. Sometimes, the structure appears to be like woven wire mesh. They appear in blow room machinery sequence in two forms – fast-revolving single cages or slow-moving condenser cages.

In the latter form, they always work in pairs. In both the forms, there is invariably a suction fan associated with it and it is made to draw air through these perforations. The ducting from the fan is led through the inside hollow of the cage.

Thus, the suction from the fan ultimately results in drawing air through the outer perforations of the rotating cage. The perforations are comparatively very small. Therefore, only an air accompanied by possibly very fine dust can enter these holes. The strength of the suction created within a cage depends upon the speed of the fan.

As mentioned earlier, this air is drawn from the beater side, and hence it controls the speed of the transportation of cotton from the beater to the cage. Along with the fine dust, sometimes, even very short fibres find their way through these small perforations and are carried right up to the fan and get deposited on a very finely meshed screen. The normal cotton lint, being larger in volume, however, does not pass through the perforations of the cage and merely gets deposited over the perforated area of the cage.

Basically, there are two types of cages used in blow room line: (i) A fast-moving, Shirley-type (Shirley wheel – Fig. 4.32) of cage and (ii) A slow-moving condenser cage (Fig. 4.31). Fast-moving cages run at much higher speed (75–85 r.p.m.) and provide a greater perforated surface area for the oncoming cotton.

The deposited layers are thus thin and allow the perforations to draw through more dust and impurities. Such types of cages are used mainly for conveying the cotton from one machine (or place) to the other. As against this, the condenser cages, as the name suggests, move comparatively very slowly (7–10 r.p.m.) and usually work in pairs. They are also perforated; but owing to their slower speed, they deposit a comparatively thicker sheet of cotton on their surface. The thickness and the mass of cotton sheet, more or less, are made to match with the desired

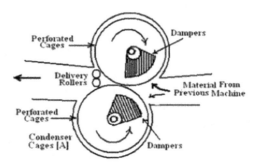

FIGURE 4.31 Condenser Cages[1,6]: As the name suggests, they are used to condense the cotton material in a sheet form. The dampers restrict the suction action over a specific area.

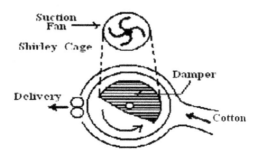

FIGURE 4.32 Fast Moving Cages or Shirley type of Cages[1,6]: They are basically used to transport the material from one machine to the next one.

dimensions and weight of the final lap. Here, the perforations do allow the sucking of fine impurities, but their proportion is comparatively small owing to thick deposition of cotton sheet.

Yet another type of cage is shown in Fig. 4.33. All the types of cages work with the help of dampers which are placed inside them. These dampers are stationary and have an opening which is directed towards the incoming path of cotton. It is through this damper opening that the suction fan draws the air from a fixed direction. The outer perforated drum rotates so that always a fresh area is brought in front of this damper opening.

The incoming cotton is continuously drawn over the cage surface. As the damper restricts the suction zone only to a specific perforated area of the cage, when the cotton sheet moves further around the perforated outer cage surface, it moves away from this suction zone. The fleece, thus, becomes free from the cage surface. In the case of condenser cage, it is usually taken ahead by a pair of cage delivery rollers; whereas, in the case of Shirley type cage, the cotton gathered on the cage, simply drops down by gravity.

Under the ideal conditions, the airflow on to the cages should be uniform across their width to ensure an even distribution of cotton. Especially, the faults like conical or barrel shape laps occur owing to lack of uniform air suction across the

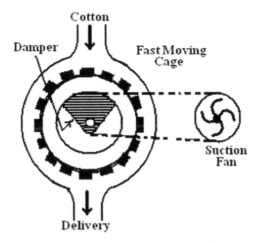

FIGURE 4.33 Another typical type of Fast-Moving Cage[1,6]: It is for conveying the material from a machine to the next machine. Conveying Cages.

width of the cage surface. Sometimes, air valves are provided to the pipes entering the central damper area and provide the necessary control over the suction.

These valves enable adjusting the correct volume of air withdrawn from each side of perforated cages The barrel shape laps are formed, usually due to excessive fan speed creating air turbulence inside the hollow cage.

Another source of trouble with condenser cage is split laps. The cotton sheets divide themselves into two halves like a sandwich. This is seen when the lap is unrolled in the next process – Carding. It is owing to approximately equal amount of cotton depositing on each cage, thus consolidating virtually two separate sheets on each cage. Even when these two sheets are combined to form one single sheet, they refuse to fully merge into each other, even after being pressed and calendered heavily. To overcome this difficulty, most of the cotton is directed onto one of the cages, preferably the top one, leaving relatively a small deposition on the bottom cage.

This is achieved by suitably adjusting the damper, air control valve and the fan speed in combination. Sometimes, the bottom cage is made of smaller diameter or top cage is slightly displaced towards the incoming cotton. Both help in offering more deposition on the top cage. Addition of soft waste exceeding 2–3% also leads to another typical defect – lap licking. This again is seen when the lap is unrolled in the next process. Here, part of the layer during unrolling, sticks to the inner layer thus leading to uneven feed to next machine – Card.

Every care should be taken to avoid the accumulation of fibres or dust in the perforations of the cage. This is to ensure more uniform distribution of cotton over its surface. A periodic brushing or cleaning of the cage from inside helps in such cases. Finally, an exhaust from the suction fan working for the cages must be suitably led into a system which would not create any backpressure. When filter bags are used, the filter cloth must be cleaned from time to time. Usually, big underground cellars are provided for letting in the exhaust air. This permanently solves the problem of backpressure. Even it helps in preventing any kind of air turbulence around the cage suction area.

4.12 FILTERS

Air filters (Fig. 4.34) are considered essential for any blow room line, whether conventional or modern blow room. They are designed to receive dusty air from the suction fans situated at various places in blow room – around stripping rollers in bale breakers/feeders/openers, or at condensers/cages used during the normal transportation of cotton. While filtering the air, they allow the return of clean air into the department. In a way, they help to maintain clean atmosphere within the department by avoiding pollution. The filters are available with the number of bags varying from two to six, the number depending upon the volume of air to be filtered. Earlier, the bags were made from cotton flannel. But the problem with these bags was the cotton flannel used to get chocked-up in short time, thus developing the backpressure. Nowadays, synthetic mesh is used for more durability. The cleaning of the pores in this case is also much easier. The smoother inside surface of synthetic mesh allows the dust, fly and trash to easily drop down into dustbins by gravity. These dustbins are divided into compartments so that while cleaning each of the bags, the suction through it can be cut-off for a short time. This allows the bag cleaning during normal working.

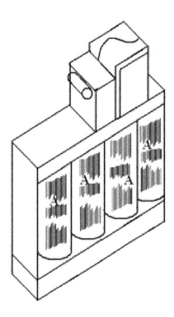

FIGURE 4.34 Air Filter Bags[1,6]: They collect the trash sucked from the cages and allow the air to pass through their perforations.

4.13   KIRSCHNER BEATER & LAP FORMING

These types of beaters have gained lot of popularity with the advent of modern blow line, though the beater itself, was available for many years and was being used in conventional blow room line. In conventional blow room line, with the introduction of improved cleaners in the early stages of blow room, the trash extraction is more effective and hence, it becomes more important to make more uniform and well-textured laps in the final stages of blow room sequence. As mentioned earlier, the scutcher is invariably the last machine used in this sequence. The Kirschner Beater (Fig. 4.35), in this respect, has been best suited for this role.

The beater consists of 3 armed spiders (A) mounted on the shaft (B). The spiders have the wooden staves (C), which are firmly fixed at their end. Each stave carries a large number of hardened and well-pointed steel pins (D) arranged in a staggered formation so as to ensure adequate combing action on the cotton fleece fed across the full width of the beater. There is a slight taper given to the wooden staves on which are mounted steel pins. Thus, the leading edge is thinner ($R_1 < R_2$) than the trailing edge (Fig. 4.36). This makes the penetration of the pins more gradual and yet more effective into the incoming fleece. It also avoids any tendency of sudden plucking-off of the tufts from the fleece.

There are three types of pin-sets – coarse, medium and fine. They are available for the use with corresponding wooden staves. The coarser pins are fewer in number but quite strong and hence are used when processing short staple trashy cotton. As against this, the fine pins are much closer and denser, but comparatively more delicate, and hence they are used for fine staple cottons containing much less trash content. The medium pins are best suited for medium staple varieties which contain moderate trash.

The beater (Fig. 4.35) is fed by a pair of feed rollers (E). However, pedal and feed roller system (F) regulates the feeding to the beater. The grid bar system (G) is

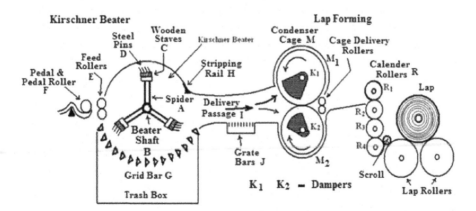

FIGURE 4.35   Kirschner Beater & Lap Forming[1,6]: A final machine in the conventional system to prepare blow room laps. These laps are processed in subsequent carding department to produce sliver.

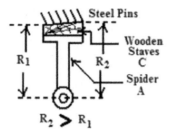

FIGURE 4.36 Tapering of Wooden staves[1,6]: It helps in gradual penetration of the needles into lap fringe.

placed under the beater and assists in cleaning the fleece. The stripping rail (H) is normally set with the beater pins at 1/16 inches. As explained earlier, the function of stripping rail is to avoid over-working of cotton. A closely set rail directs the cotton quickly into delivery passage (I). The cotton, on its way, passes over the grate bars (J), which allow some fine dust and impurities to pass through them. Owing to suction at the cages, the cotton is finally deposited onto the pair of condenser cages (M_1 and M_2). The dampers (K_1 and K_2) play their usual role of restricting the oncoming cotton to a specific area over the cage. Thus, a comparatively thick sheet almost equivalent to lap thickness is formed on the cage surface.

In the conventional blow room, the beater was usually run at 750–800 r.p.m., though sometimes little higher speeds were also used. The beater has comparatively mild cleaning action, but gives very intensive opening of the tufts. Therefore, it may not be considered as good cleaning agent. However, owing to its very effective opening action (due to fine pins), it produces very fine tufts, which are essential for producing a smooth, uniform and well-textured sheet of cotton, called lap which is free from stringiness. A lap possessing these qualities is a pre-requisite to good carding action, and hence this beater was considered to be an indispensable part of all conventional blow room line producing laps. In typical modern blow room lines, its versions are suitably included. In conventional line, its place was always fixed, being the last machine in the blow room sequence. Finally, the cotton is delivered by cage delivery rollers and the lap is subjected to a very heavy calendering action.

4.14 FEED REGULATION

A pair of feed rollers or pedal and feed roller is basically used to feed the various beaters, including Bladed Beaters and Kirschner Beater. These systems are used when the cotton is gripped by the feed system and the issuing material is struck by the blades or pins. Obviously the striking action of the beater blades is much stronger. However, in modern blow room line, in many instances, the cotton is acted upon by the blades or pins when it is in free-flow condition.

Even when the feeding is ultimately done by pair of feed rollers, the feed regulation is done by Pedal and Feed Roller. In this case (Fig. 4.37), the pedal is supported on the sharp-edged fulcrum and carries a long pedal tail on the other side. The pedal nose is specially curved to suit the contour of the pedal roller. The tail is

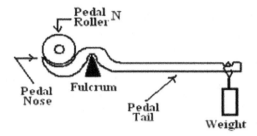

FIGURE 4.37 Feed regulation[1,6]: through pedal system. It controls the amount of feed to the beater.

weighted by using a hanging weight. The distance across the machine is covered by many such small pedals, placed side by side, and each is weighted in a similar fashion. The weights put the necessary pressure on the pedals from underside. This pressure thus acts on the upper feed roller and is responsible for creating a grip between upper pedal roller and the curved bottom pedals. Thus, the pedals are always in contact with the pedal roller. However, over all these pedals, there is only a single pedal roller.

When the sheet of cotton passes through pedal roller and pedals, the latter are pressed down depending upon the thickness of the sheet passing through. This raises the pedal tails against the weight. With a normal thickness of lap sheet passing through the pedals and roller, the pedals are pressed down to a definite distance.

With the normal lap thickness, the belt on the cone drums B and C also remains in the central position (Fig. 4.39). As can be seen from Fig. 4.37, the drive from cone drum to the pedal roller is reached through a shaft D, worm E and worm wheel F. The clutch G is operated from the starting handle (old scutcher), positioned in front of the machine. Thus, the drive to the pedal roller can be disengaged whenever the process of lap making is required to be stopped. In the normal running position, however, the clutch is engaged and the pedal roller directly receives its motion from the cone drums.

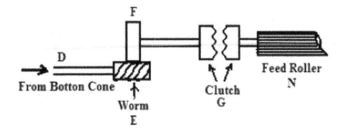

FIGURE 4.38 Drive to Pedal Roller[1,6]: A provision is made to clutch or de-clutch the drive, when the feed is required to be stopped.

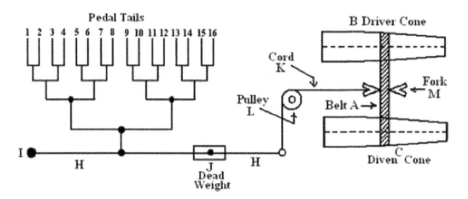

FIGURE 4.39 Pedal Feed Regulating Motion[1,6]: The resultant movement of the pedals is finally reached to the belt shifting mechanism to alter the speed of pedal roller.

The cone B (Fig. 4.39) is a driver cone, whereas cone C is driven cone. The drive to the pedal roller is given from this driven cone C. There are in all 16 pedals to cover the entire width of the lap sheet (Fig. 4.39). When a thicker portion of the lap (thicker than normal) passes through pedal and pedal roller, the corresponding pedal/s are pressed down, thus lifting their tails. This lifting results in raising the lever H around the fulcrum I against the moment of the dead weight J.

The raising is finally conveyed through cord K, passing around the guide pulley L to shift the position of the belt fork M. The belt fork shifts the belt to the smaller diameter of the driver cone B, and hence the speed of the driven cone C and that of pedal roller is reduced. Ultimately, the heavier material is fed to the beater at a slower speed. This helps in maintaining the mass of the material fed to the beater more or less constant.

It is necessary that the pedal tails are allowed to swing up or down freely whenever the pedals are pressed or raised owing to thicker or thinner portion of the

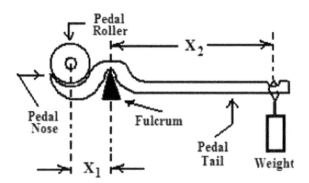

FIGURE 4.40 Enlargement of movement of Pedal Tails[1,6]: The magnification facilitates sizable movement of the tail and leads to noticeable change in driven cone speed.

cotton sheet. For this, a periodic maintenance is required to clean and lubricate the edge over which the pedals are made to swing. Further, a correct variation in the speed of the pedal roller according to the thickness of the cotton sheet is equally important factor to bring about the necessary change in the speed.

For this, the contour of the drum is specially designed to suit this purpose. Similarly, the sensitivity of shifting the belt is also required to be sufficiently higher. A long tail (Fig. 4.40) helps in magnifying the small resultant movement of all the pedals and helps in improving the sensitivity. This is necessary because; as soon as the heavy or light portion of the lap is sensed by the pedals, the appropriate speed alteration should take place without any delay.

4.15   CONTOUR OF THE CONE DRUM

The change in the speed of the cone drum and hence that of feed rollers has to be inversely proportional to the thickness of cotton sheet passing through the feed system. Thus, if it is assumed that the thickness is doubled, the speed of the feed roller should be halved. Suppose, AB is the length of cotton sheet passing through the pedal roller and pedal during a certain time, with a normal thickness 't' (Fig. 4.41), then doubling this thickness would necessitate that half the sheet should be delivered and hence point (C) in the figure. The points D, E F and G are for 1/3th, 1/4th, 1/5th and 1/6th length of the sheet required when the thickness is 3, 4, 5 and 6 times, respectively. On joining these points, a hyperbola is obtained.

Thus, it can be seen that the relation between the pedal roller speed and the thickness of the lap sheet is hyperbolic. While designing the contour of the cone drums, a certain limit of extreme speeds must be assumed. This is because, in actual practice, the lap thickness would not vary to the extent – six times. Therefore, if minimum to maximum speed ratio is (say) 1:2, the driven cone drum speeds can be correspondingly found. If the driver cone drum runs at 100 r.p.m. and if the maximum and minimum diameters of the cones are 8 inches and 4 inches, respectively, then the speed of the driven cone would range between 200 r.p.m. and 50 r.p.m.

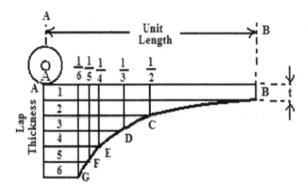

FIGURE 4.41   Contour of Cone Drums[1,6]: The hyperbolic surface of cone drum ensures that the change in the speed of feed is inversely proportional to the thickness of the sheet fed[1,6].

TABLE 4.1
Diameters of the Cone Drums

Driver Cone D″	Driven Cone d″	Speed Ratio	Speed of D r.p.m.	Speed of D r.p.m.
8.0	4.0	2.00	100	200
6.86	4.14	1.65	100	133.3
6.0	6.0	1.00	100	100
4.8	7.2	0.66	100	66.6
4.0	8.0	0.50	100	50

Also, when position of the driver and driven cone drums is fixed, their centre-to-centre distance will remain constant (Fig. 4.42). Therefore, as the same belt runs over the two cone drums, it will automatically lead to the following expression:

Diameter of DriverCone + Diameter of DrivenCone = 8 + 4 = 12″ = Constant

The respective diameters for attaining certain speeds of driven cone drum in the above case are shown in the following table. It may be noted here that the diameters assumed in the table briefly give an account of the related speeds attained by the driven cones. However, while drawing the contour, many more values in between the two extremes need to be found-out, so that a smooth contour curve can be drawn.

The length of the cone drums are decided by both the movement of the belt and its width. The pedal movements decide the distance through which the cone belt shifts; whereas the power transmission decides the width of the belt. However, a certain minimum workable belt-width is chosen so that the belt does not slip over the drum and, at the same time, brings about the effective changes in the speed of driven cone at the time of belt shifting. The ratio of the two portions of the pedal around the fulcrum (Fig. 4.40) i.e. $X_2 : X_1$ is normally more than 5. This

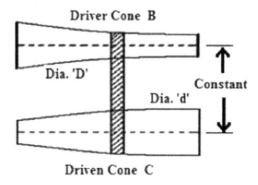

FIGURE 4.42 Driver & Driven Cone[1.6]: A constant distance allows the same belt to run at all its positions on cone.

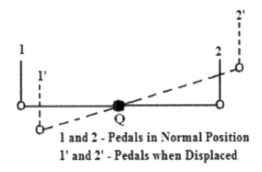

1 and 2 - Pedals in Normal Position
1' and 2' - Pedals when Displaced

FIGURE 4.43   Pedal Tail Displacement[1,6]: The average of the movements of all tails is finally used for belt shifting.

substantially multiplies the initial small movement of the pedal nose to cause more effective shifting of the belt.

When the lap is not uniform across its width, the displacement of the corresponding pedals would be in either direction. Some pedals would be pressed down for heavy portions, whereas, some others would be raised owing to thinner portions. After summing up these movements, only a net displacement is conveyed for belt shifting (Fig. 4.43).

In an extreme hypothetical case, suppose half the pedals are pressed down, whereas the remaining half are lifted-up, the distance being equal and opposite (as shown in Fig. 4.43), the centre Q would remain undisturbed at its original position. Thus, there would no indication passed-on for any change in the position of the belt. Had the pedal been extending full width across the feed roller and not split-up in small sections (Fig. 4.44), the displacement of the pedals would have been biased for bigger tufts only. By having more number of pedals, each of them is able to detect its own section of lap sheet of approximately 2 inches and thus takes care of unevenness of small-width sections across the lap sheet. It can be seen from Fig. 4.39 that there are 16 such pedals covering the entire width of the lap and each judges its own quota of lap width quite independently. The Tripod and Lever-Bowl types of conventional mechanical feed regulating motions are shown in Fig. 4.45 & 4.46.

It can also be seen (Fig. 4.40) that the linkage for each pedal tail enlarges this pedal movement and subsequently passes it further.

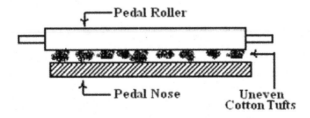

FIGURE 4.44   Uneven Material Under Pedals[1,6]: This problem is counterbalanced by using large number of pedals and making the width of each smaller.

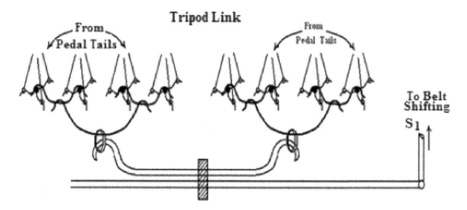

FIGURE 4.45   Tripod Regulating Mechanism[1,6]: Each tripod is connected to three corresponding tails of pedals. Thus, total tripod decide the number of pedals used.

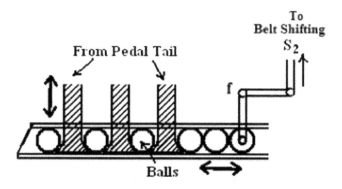

FIGURE 4.46   Lever & Bowl Feed Regulating Mechanism[1,6]: The movement of the tails directly lift the tapered levers which displace the bowls to lead to shifting of belt on cone drums.

The sum total of the net movement is then conveyed to the belt fork for bringing about the speed change, if necessary.

It may be mentioned here that, with the introduction of variable speed servo motors, the mechanism of feed control has become much simpler and far more instantaneous in its action in relaying the changes in the lap thickness. In such a case, it is not necessary to maintain such an elaborate mechanism of conventional feed-regulating mechanism. This is because there are many improved thickness sensing devices which directly relay the sensing to these servo motors.

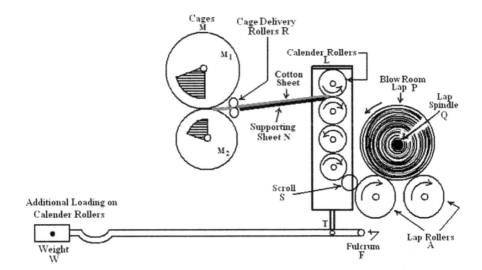

FIGURE 4.47 Lap Formation at Scutcher[1,6]: It is important to form a very compact lap so that it stands intact during its transportation to next department – Carding.

4.16   LAP FORMATION

As explained earlier, the sheet of cotton coming from the cage M (Fig. 4.35 & 4.47) is drawn forward by cage delivery rollers. The cotton sheet then passes over the supporting metal sheet N and is led on to the calendaring system.

There are four calender rollers placed one above another. They are very heavy and are made of highly polished steel. The rollers are suitably placed in vertical slides and in addition to their own weight are heavily weighted. In old scutcher, this weighting was obtained by mechanical means, where the weight W (Fig. 4.47), on either side of the machine was carried by a long lever fulcrumed at F. The connection at T finally relayed the additional weighting on calender rollers quite effectively. In modern versions of conventional scutchers, the pneumatic pressure was used for this purpose.

On its way to the lap rollers-A (Fig. 4.47), the lap sheet passes under the scroll S, which again is a very heavy roller. The surface of the scroll roller is similar to a threaded screw, and hence it causes indentation marks on the sheet. This helps in easy unrolling of the sheet in the next process and reduces the tendency for the adjoining layers to stick to each other. The lap spindle Q on which the lap is wound rests on the lap rollers (A) and is rotated by its surface contact with them.

4.16.1   Pressure on the Lap Spindle

The cotton sheet coming through calender rollers is finally tucked around the lap spindle and this initiates lap formation. In conventional blow room, the pressure is applied on both the sides of the spindle by means of racks (Fig. 4.48). The pinion B engaging the rack A is geared through wheels – C and D to the brake drum shaft E.

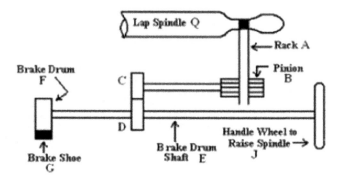

FIGURE 4.48 Lap Spindle Loading[1,6]: On this spindle a lap is wound. For making compact, a brake shoe is provided and it does not allow the spindle to rise so easily with growing lap diameter.

The brake drum F has a brake shoe G, which is in constant contact with the drum, thus applying brake to free rising of the lap. When lap sheet is wound on the spindle, it slowly grows in diameter.

The lap spindle thus has a tendency to rise-up on the lap rollers. A heavy weight carried by a weight lever H (Fig. 4.49) puts the pressure on shoe G and creates the braking pressure on the drum. The friction between the shoe G and the brake drum F does not easily allow the drum to rotate. However, when the material is wound on the lap spindle, it creates tendency for the spindle to rise-up. When this tendency overcomes the braking pressure between the shoe and the brake drum, only then is the spindle allowed to rise up.

This makes a more compact lap and does not allow it to swell unnecessarily in diameter. The pneumatic devices have been developed in later years for applying the pressure on the lap spindle during lap building. The compressed air, at a pressure of 3.72 kg/sq. cm or (50 lbs/sq. inch), is fed to a cylinder in which a piston (Fig. 4.50) is made to operate. The piston rod is extended to form a toothed rack, which engages a pinion. The connections to the lap spindle, thereafter, are similar. The pinion is allowed to rotate a little at every time, because of the building of diameter of the lap. However, this pressure has to overcome the air pressure. An air in-let is provided with a reducing valve and the pressure gauge to control the piston pressure.

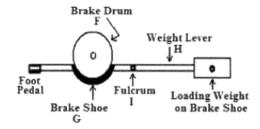

FIGURE 4.49 Brake Drum & Shoe[1,6]: It restricts easy rising-up of the lap.

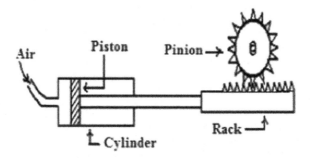

FIGURE 4.50 Pneumatic Pressure on Rack[1,6]: In this, mechanism, the air pressure is used to restrict an easy rising-up of lap spindle, when the lap is being built-up.

A release valve is also provided to maintain a constant pressure on piston throughout the building up of lap.

4.16.2 Lap Length Measuring Motion

In old Scutcher, the full-length stop motion was mechanical (Fig. 4.51). A worm (B) on the bottom calendar roller (A) is made to gear with worm wheel (C) on the side of the shaft (D). The change wheel (E) at the end of this shaft drives the knock-off wheel (F).

After a certain number of revolutions of calender roller, the wheel (F) is designed to make one revolution. After each revolution of wheel F, the knock-off bracket (G) placed on this wheel, is set to knock-off the starting handle lever, thus stopping the

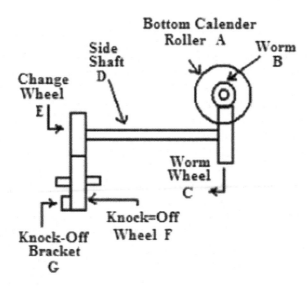

FIGURE 4.51 Full Length Stop Motion[1,6]: With conventional blow room, the machine was stopped after making a precised length of each lap.

lap rollers and the feed. One complete revolution of knock-off wheel (F) is equivalent to certain number of revolutions of bottom calender roller, which in turn, decided a fixed length of lap delivered. Here, the lap doffing with old scutchers was then done manually by the tenter or operative.

After each knock-off, the racks are first raised. The full lap with the spindle inside is removed. The lap rods are put inside the hollow lap spindle. The spindle is pulled from the other side, thus keeping the lap rod inside lap. The spindle is put back on lap rollers. The pressure racks are again brought down on the spindle necks to continue exerting pressure on the lap. The machine is started. The lap delivered by calender roller is once again tucked around the spindle to initiate the next lap building.

In subsequent versions of scutchers, the length of the lap was arranged by adjusting lap length counter meter. The lap doffing was done automatically by using pneumatic pressure. The replacement of the spindle and repositioning of the racks were all done automatically. During all these operations, there was no knock-off and the machine continued to deliver the lap. Therefore, while removing any human involvement, it increased productivity.

4.16.3 BLOW ROOM MACHINERY SEQUENCE

In conventional blow room line, the selection of textile machinery in the blow room is vastly influenced by the type of cotton to be processed. A short staple cotton needs a longer sequence because of higher trash content; whereas, a long staple cotton, which is fine, delicate and less trashy, requires comparatively shorter sequence. When processing man-made fibres, the machines involved are quite different. There is no trash with man-made fibres and what is needed with these bales is a thorough opening of the matted mass.

The sequence of machines arranged in any conventional blow room line has to be such that it includes cleaning units as well as opening units. The total number of machines employed for any type of work generally varies from minimum 2 to 7. When such a line is installed, a spinner usually has a lot of choice in selecting the appropriate number of machines in the sequence to suit the requirement. In certain cases, some machines are required to be omitted, especially when processing superior varieties of cotton. The total sequence, therefore, needs to have some provisions to by-pass certain machines.

Earlier, there was a practice of giving intense blow room treatment for achieving maximum possible cleaning. This was done by selecting, on priority, those machines, which would give very harsh action on the material processed. This, however, led to a huge blow room waste containing much higher proportion of lint. Thus, it was a serious two-fold problem. On one hand, the crude cleaning methods of those typical beaters caused severe fibre damage, which ultimately led to higher neps, more yarn irregularity and lower yarn strength. On the other hand, it also led to poor yarn realization, owing to higher lint loss. With conventional beaters in the old blow room sequence, there was another problem. As more machines were required to reach a certain cleaning efficiency level, the sequence became unnecessarily longer, occupied a larger space and involved higher maintenance. The use of mechanical means of harvesting further aggravated the situation, particularly when better grade cotton contained more trash. Selecting blow room sequence in such cases became a real crucial problem.

4.17   SINGLE PROCESSING

With improved cleaning techniques in machines and with modified approach in the later versions of conventional blow room line, the effectiveness of both the opening and simultaneous cleaning considerably reduced the number of cleaning units. The use of single process blow room line made the process quicker, shorter and economical. It is sometimes better to have a stronger, uniform and nep-free yarn at the cost of a little lower cleaning efficiency at blow room than over-working the cotton to attain a more desirable cleaning level. In this respect, the earlier concept of carrying-out more cleaning at blow room to reduce the load on card does not hold true. This is because with the modern trends, it is far more advisable to let the blow room work at its normal cleaning capacity. Intensifying blow room treatment not only involves higher blow room droppings but also leads to more nep generation.

As regards the conventional cards, their ability of completing the unfinished cleaning job of blow room was limited. It was also thought that the basic function of card being individualization of fibres, loading them additionally for cleaning would jeopardize their main function. In addition, this extra load of cleaning would have caused more and faster wearing of the card wires, especially of cylinder and flat. The present generation cards, however, are far more capable of carrying out this additional function of cleaning – the unfinished job. Therefore, avoiding unnecessary intensified blow room treatment has three distinct advantages: (1) Blow room loss is kept within tolerable and reasonable limits. (2) The possible fibre damage and excessive nep generation is effectively restricted. (3) This is also likely to result into better yarn properties, especially tensile properties.

4.17.1   Features of Conventional Single Process Blow Room Line

Single Processing was one of the most important features of improved version of conventional blow room line. The various machinery sections forming the old blow room were merged into one and this led to a single line working continuously and equipped with some useful modern machines which were subsequently introduced. Some typical features are given below;

4.17.1.1   Single Processing

There is no man-handling of material in between. Once the bale cotton is fed to a pair of Bale Breakers or multiple Blenders, the material continuously travels ahead till the final lap comes out through automatic doffing device at scutcher.

This avoids any possible irregularity in the first few yards of every lap sheet. Further, in the old system, where there was manual doffing and the tenter was also lazy, he used to allow the full lap to roll on the lap rollers. This led to stretching of the outer layers of the lap. With auto-doffing, when the full lap is complete, it is detached and simply rolled out of running lap rollers.

4.17.1.2 Automatic Doffing

The introduction of fully automatic doffing has eliminated any possibility of non-uniform and spoiled laps. As the scutcher is not required to be stopped at the end of each full lap, it also helped in increasing the production.

4.17.1.3 Economical

Very old conventional blow room line consisted of four sub-sections – pre-mixing machinery set-up; mixing room; blow room sequence up to Breaker Scutcher and Finisher Scutcher. In a continuous single processing, these four are merged into one. This not only has reduced a lot of floor space but has enabled the use of much less labour to mind the whole blow room line. Even the power consumption is comparatively much less.

4.17.1.4 Good Blending

Though the number of doublings in later version of conventional blow room process (single processing) was reduced considerably, the introduction of blenders and auto-mixer gave very efficient and homogeneous mixing of various lots of cotton used in a given mixing. In modern blow room, there are far superior blending techniques (machines like Integrated Trunk Mixers, Volume Mixers, Tuft Blenders), which have amply compensated these aspects of satisfactory blending.

4.17.1.5 New Beaters

The conventional single process line included beaters like Step Cleaner, Axi-Flow, Air Stream Cleaner, and Shirley Opener. All of them incorporated a very effective opening and cleaning action and reduced the trash content in the final lap quite appreciably. This enabled considerable reduction in number of machines in a blow room sequence. It also led to less fibre damage. It was because these beaters helped in avoiding harsh beating and yet opened the cotton thoroughly. This led to an improved cleaning efficiency and effective extraction of trash.

4.17.1.6 Automatic Feed Regulation

In single-process blow room line, a continuous flow of cotton is always maintained. The machine utilization is thus improved. At the same time, the machines are never over-loaded and hence it leads to higher cleaning efficiency. The feed regulating motions used at various places in the sequence control the unnecessary piling-up of stock during the processing and make it possible to save a large amount of labour and floor space.

The main function of the feed regulation is to give uniform feeding and maintain a certain uniform rate of movement of the material through the machine. This helps in improving the regularity of the final product (lap). In modern blow room installations, however, the chute feed to card has obviated any necessity of producing laps.

The use of electronic devices has become a part and parcel of the control systems. In that, the use of micro-switches, photocell, relays, and other electronics has made the controls very effective. The fault detection has also been made easy. In place of old styled cone drums, servo-motors are introduced for speed variation and the control over the feeding has become more precise, accurate and instantaneous.

4.17.1.7 Trash Content

Introduction of new types of beaters like Axi Flow, ERM Cleaner, Mono Cylinder Cleaner, Air-Stream Cleaner has resulted in less number of machines required in a sequence. Even then, the cleaning efficiency of the overall sequence has been much improved. With lower trash in the material delivered and more uniform laps, the load on card is greatly reduced.

4.17.1.8 By-Passing Arrangements

Many by-passing arrangements are provided in a single process line. These allow such machines which give more intensified action in the sequence to be omitted for a particular mixing. With such by-passing arrangements, it is possible to omit certain machines and lead the material directly to the subsequent machine. giving less harsh action. The micro-switches are provided to ensure the correct path so that unless and until all the appropriate flap doors are secured in their proper positions, the sequence does not start.

4.18 ACTION OF VARIOUS CONVENTIONAL MACHINES

As the important object of the blow room beaters is to remove the trash by opening the matted mass of cotton, it is interesting to compare the action of various opening and cleaning machines used in blow room. The action of the various beater and openers is analysed in the following discussion.

1. **Bale Breaker and Hopper Feeder** hold the cotton on the spikes. It is spike to spike action that reduces the mass of cotton from matted condition to somewhat opened tufts. Here the leather flap beater gives a mild beating action. The leather flaps, not only strip-off the cotton from the spikes, but also strike it gently against the grid bars situated below. It results in removing some of the impurities. The beaters like **Step Cleaner, Axi Flow** are developed on these lines. The cotton in these machines is also struck by the striker blades but it is not gripped by any gripping device. The cotton in free flow receives beating by the beater blades but this beating is much gentle. This action of blades results into opening as well as removing some trash from the cotton mass.

2. The beaters like **Crighten** and **Axi Flow** strike the cotton when it is carried by the air currents. However, as the cotton has to move up against the gravity in Crighten, its action on cotton is very harsh. In any modern blow room line, therefore, Crighten has been totally discarded. In Axi-Flow, the cotton travels horizontally through the machine and the beater is far more versatile. The spacing of the grid bars, their angle and setting with respect to beater

determine the cleaning efficiency. The strength of the air currents should, however, be controlled. It should neither be too strong so as to disturb the beater functioning, nor too weak to allow the cotton to be over-treated. These beaters are very effective in removing the trash from the cotton. However, like Crighten, the beater rotations in Axi Flow do not help the cotton to move ahead through the beater. A little higher lint loss in this case is, however, unavoidable.

3. **Bladed Beater:** The beaters in this category strike the cotton when it is firmly held by the feeding system. The distance between the point of issue of the material and the beater blades becomes very important and should always be a little more than staple length of fibres to avoid any fibre damage. The damage can still occur, if either the construction or the speed of the beater is unsuitable for the fibres being processed.

The speed of the beater, together with the number of blades, decides total number of blows received by cotton sheet. If, however, the length of the cotton sheet fed during the same time is changed, the effectiveness of beater blows –'Beats per inch' – changes. Altering the beats per inch has a direct effect on the cleaning (also opening) action of the beater. In this respect, though the construction of the two- or three-bladed beaters is similar, the extra blade that the latter has, increases the beats per inch. For example, if the beater speed is 800 r.p.m. in both the cases and the delivery rate of feed roller, in each system, is 40 inches per minute then,

Three Bladed Beater = (800 × 3 blades)/40 = 60 beats per inch

Two Bladed Beater = (800 × 2 blades)/40 = 40 beats per inch

With higher beats per inch, the action of beater becomes more powerful in striking the lap sheet and is thus more effective in removing the trash entrapped in cotton tufts. This is because the size of the tufts in this case is smaller. However, the force with which each beater blade strikes on the fringe issued from feed roller is greater in the case of two-bladed beater as compared to three-bladed beater. Hence the effectiveness of the action of two-bladed beater is still comparable with that of three-bladed beater. Three-bladed beater, however, has a slight edge over two-bladed beater because the beats per inch can be varied over a wider range. There is also less wear and tear of moving parts owing to better balancing of beater.

4. The **Porcupine Opener** has a very effective combing action owing to smaller size of the striker blades, but the sharp square-edged striker blades also give forceful blows to cause a good cleaning action. The striker blades are off-set in such a manner as to strike the cotton across the full width of the machine in every revolution. Out of the several off-set striker blades mounted on each disc, there are different pairs of striker blades which exactly follow the same path around the beater in every revolution. Hence, the beats per inch are comparatively much less. [For 960 r.p.m of beater and 130 inches per minute

– 14.7 beats per inch]. Therefore, while processing longer staple cotton, Porcupine Opener does not show any tendency to over-beat the cotton owing to lower beats per inch. This is also due to the fact that the width of each blade striking the fringe of cotton is very small. Thus, the force with which each blade strikes the fringe results in pulling away only a very small width of lap fringe. This avoids any possible damage to the fibres in the fringe.

5. **Kirschner Beater** in this respect is a very good opener. This is because its fine pins give more of a combing action on cotton sheet than any serious beating. The penetrating effect of the pins helps in opening the tufts to a very fine size. However, this does not help in much of cleaning action, as the smaller pins do not impart any forceful blows to jerk-out the impurities when the tufts are thrown on to the grid bars. Whatever cleaning that results is merely because of its opening action. However, when processing man-made fibres, Kirschner (or its type) Beater is very useful and effective due to powerful opening action of its pins.

6. The opening power of **SRRL** is unique. The vertically positioned saw tooth cylinders, each of 12 inches diameter, are spaced at a very fine gauge (1/8 inches) and all the cylinders run around 400 r.p.m. The metallic wires on these cylinders and that on the doffing cylinders, comb-out the cotton tufts very thoroughly. The size of the tufts picked by each of the fine saw teeth is very small. However, the whirling and rolling action of cylinders is prone to slightly higher nep generation, especially when cotton with lower maturity is processed. The success of the use of saw tooth type of metallic wires in blow room depends mainly on the correct positioning of these beaters in the sequence. If such beaters are placed too early in the blow room sequence, then more often than not, the wires get damaged owing to their mounting procedure and poor strength when handling matted mass. However, when these beaters are placed in the middle or in the later part of blow room sequence, the saw-tooth wires offer very good opening and also help in giving satisfactory cleaning. This improves both the cleaning efficiency and lap regularity, the latter due to small tuft size.

7. The action of **Shirley Opener** or **Air Stream Cleaner** is very effective in aerodynamically separating lint and trash. The action on the cotton during cleaning is never harsh. In these beaters, the principle of difference in buoyancies of lint and trash helps in removing the latter. In both saw tooth wires are used.

4.19   PROCESSING OF MAN-MADE FIBRE & BLENDS

The need for using man-made fibres is basically due their unique properties such as strength, extensibility, crease resistance, wrinkle recovery and wash and wear properties. The fibres like viscose and polyester are commonly used in cotton textile industries in their single or blended form. Some popular combinations are:

Cotton/Viscose: Viscose improves uniformity and appearance. Also, it is used to reduce the cost of mixing.

Polyester/Cotton: Polyester gives wash and wear properties. It improves crease resistance, durability, strength, while cotton provides comfort and reduces static.

The processing of man-made fibres can be done in two different ways: first when these two components are separately processed up to a certain point (say card) and then combined together, and second, when they are blended in blow room itself. Many machinery manufacturers like M/s Trutzschler and Rieter have come out with a blow room line specially suited for processing 100% man-made fibres.

Being free from impurities, usually the blow room sequence of machinery for processing man-made fibres is much shorter. The main focus is to open-out the matted mass packed in the bale form. It is customary to feed the material initially to a machine carrying out similar functions like Hopper Feeder. It then passes through two Kirschner Beaters, which are specially accommodated on Scutcher. Here, the traditionally used two or three-bladed beater is replaced with yet another Kirschner type beater. Replacing bladed beater by Kirschner is obviously needed for more opening action required for man-made fibres. This is because they are in quite a matted and compact condition when the bales are opened in blow room. With absolutely no trash, cleaning at any place during blow room treatment is not intended. Even then, the extraction of a small percentage of fused fibres is desirable.

During processing in blow room, however, some modifications are necessary. At scutcher, the reserve box capacity has to be increased to accommodate fibres with higher bulk. The speeds of inclined lattices need to be decreased so as to avoid excessive buffeting action, which otherwise, may lead to nep generation.

The weighting on the calendar rollers at scutcher is required to be increased by approximately 25–30% to get more compact laps. As against this, the weighting on lap spindle racks should be reduced to avoid lap licking. The use of scroll roller and felting finger also helps considerably in this case. The roving strands can be inserted in the lap layers at calendar rollers for effective separation of the lap-layers at card. The length and the total weight of total lap as well as the weight per unit length of lap sheet are required to be considerably reduced. This is because the higher bulk of the fibres (with polyester) would otherwise make too bulky laps. The dampers on the cages are adjusted in such a manner that most of the fibres get deposited on top cage. This considerably controls the lap licking. Excessive fan speeds are avoided. The blunt beater blades, rough edges of grid bars and roughness of inner surfaces of conveying pipes generally lead to formation of neps. A careful attention is required here to tackle this problem. Also, it is advisable to shorten the length of the conveying pipes.

When the mixing of man-made fibre and cotton is done in blow room, it may become necessary to clean the cotton first before blending it with man-made fibres. For this, the cotton is separately processed up to 'Combing'[III] and then brought back to blow room in sliver form for blending with man-made component. As the bulk density of cotton and man-fibres (especially Polyester) differs, gravimetric blending must be followed.

The sequence of machinery for blends is the same as that used for 100% man-made fibres processing. Few mills process the cotton separately up to card and bring

---

[III] In Combing, the short fibres are fractionated (removed) preferentially and this up-grades the quality of cotton.

back the carded sliver for blending it with man-made fibres in blow room. This may be practised only when the cleaning level of the material at the card is assured.

In the mill, when different types of man-made fibres or different blend ratios or even different deniers are used, it is customary to tint either the man-made or cotton component. This helps in identifying the composition or blend and avoids any mixing of these in the subsequent process. The tinting, however, is done by taking a very small percentage of one of the component fibres (less than 1.5%). This quantity is separately dyed and fully dried before its use. The tinted material is then spread in such a manner that it gets distributed over all the quantity of component taken at a time.

4.20   CONTROL OF BLOW ROOM LAP WEIGHT

The basic causes of wide fluctuations in full lap weight are:

1. Fluctuations in atmospheric conditions
2. Variation in the density of the material fed and therefore delivered
3. Inadequate blending
4. Improper synchronization of machines in sequence
5. Addition of uncontrolled percentage of soft waste

As the cotton is handled in loose form in blow room, the lap weights are easily affected by the fluctuations in relative humidity (R.H.). A large volume of air is sucked by the cages and this makes the departmental atmospheric conditions unstable. Thus, the internal atmospheric conditions are susceptible to frequent changes, especially in the monsoon season. Therefore, if the internal R.H. is not taken into account, the changes made to adjust lap weight will result in false recording of weights. This is because when the relative humidity in the department changes, the corrections made to adjust the lap weights would prove to be unjustified. The suitable corrections charts are available and they would indicate the corresponding lap weight related to existing relative humidity in the department. Sometimes a 'standard lap' is prepared and all other laps are weighed against it. The changes in the atmospheric conditions are automatically registered in this 'standard lap'. This avoids unnecessary and untimely corrections.

When the bales of different densities are used in the mixing, it usually reflects in the lap weight variation. The ideal way of overcoming this will be to mark these bales in equal parts before opening the tying-bands. This helps in ensuring their uniform feeding.

With battery of Hopper Blenders used for mixing the lots, it is always advisable to check whether their delivery rates are in conformity with the mixing proportion. The use of at least two Hopper Feeders in the line helps to ensure satisfactory homogeneity of blends and control over the feeding. The degree of opening received by the cotton in earlier sequence, in this respect, is also important.

Addition of soft waste is yet another factor governing the control over lap weight variation. Normally the addition of soft waste should never exceed 2.0–2.5%. The use of separate waste blender, in this case, is very helpful. Sometimes the rejected laps are immediately fed as and when they are received. If there is a proper system to redistribute these laps uniformly and proportionately over a longer period, it would be more beneficial.

1. **One-Metre Wrapping of Lap:** Incorrect functioning of pedals of feed regulating mechanism, irregular feeding to finisher Scutcher and fluctuations in the height of cotton in the reserve box prior to bladed beaters are generally the main causes for variation in weight per unit length of lap.

   The occurrence of heavy lap-tails is due to stopping of feed and cages simultaneously for any reason. In this case, the cotton accumulates on cages and forms thick deposition. With automatic doffing, this is greatly reduced. In addition, the first few metres which are wound on the lap spindle don't receive the effect of tension draft between the lap spindle and bottom calendar roller, leading to slightly heavier portion of lap being wound. This problem to a great extent is solved by bringing the spindle rack quickly down. The mechanical tucking of front end of fresh lap around the lap spindle also serves to avoid any further delay in this regard.

2. **Length-Wise and Width-Wise Variation:** The short-term variation along the length of lap mainly occurs owing to fault in the cage system. The deposition of cotton on the cages is affected by 'Eddy Currents'. These currents are generated due to the blades of the beaters functioning as fan blades. Even the concentricity of the cage may cause fluctuations in the lap weight. The gears which are worn out or loosely meshed or even loose bearings in their housing may affect the short-term variations considerably. With longer laps, the width-wise variation in the lap becomes prominent and hence they bulge at the middle or at the sides. This is mainly due to incorrect cage damper setting, which is responsible to carry the cotton from the beater on to the leading to cage. Even an insufficient air suction created by the fan may be the reason. In modern blow room, an arrangement is provided for obtaining separate air passage to the cages. It is controlled and distributed uniformly along the width to ensure even distribution of suction on the cage surface.

4.21   PREVENTIVE MAINTENANCE

With three shifts working, the maintenance of the machines becomes very important. Usually, the maintenance staff works in the first shift only and hence its routine has to be timed accordingly. A regular maintenance becomes more essential in single process line so as to avoid any serious break-down, otherwise it causes considerable loss of production. Any major break-down in the blow room line seriously affects not only the productivity of blow room but also hampers the working of carding section, which heavily depends upon the supply of regular feed material (in the case of both lap feed and chute feed).

The maintenance may be broadly divided into – cleaning, servicing, lubrication and replacement of worn-out parts. Cleaning and servicing must be done at the start or at the end of each shift. The oiler, during this time, should take the opportunity to grease some of the parts by using grease-gun. It is a good practice to record the break-downs and replacement of parts machine-wise. It is also important to record any repair work individually. This creates history for each machine and enables observations of some of the troublesome locations which fail frequently. When the machine is taken for overhauling, these areas can be looked into in more details to find the root cause for such failures. Also, the spares can be managed accordingly in stock and yet, unnecessary stock of some other infrequently required items can be stopped at the same time. Even then, the points mentioned below with respect to conventional blow room set-up require a routine inspection and attendance:

1. Checking the lattice – for damage to wooden lags, slackness of lattice belt
2. Oiling and greasing of all rotating parts
3. Removal of waste from the beater chambers
4. Cleaning of beater chambers, cages and grid bars – adjustment for dampers, flue-pipes to and from suction fan
5. Feed-regulating mechanism – inspection of knife rail on which pedals swing. Lubrication at this place for the pedals and at all the moving links in feed regulating mechanism
6. Sharpening and re-polishing of beater blades and grid bars
7. Checking of important settings like distance between beater blades and grid bars, beater blades and feed system and distance between beater blades and stripping rail or baffle plates, and making the necessary changes
8. Speeds of beaters and fans – a routine check-up for their measurements and corrective action in the case of any discrepancies
9. Safety devices and their up-keeping

4.22 TYPICAL SINGLE PROCESS LINES

Having discussed some typical blow room cleaning machines, it is possible to get a fair idea of how each machine is planned in a blow room machinery sequence and what is their position in the sequence. Equally possible is to imagine the treatment that cotton would get as it travels through this sequence.

In a single processing, all the machines are arranged in a certain sequence. The position of each is chosen on the basis of what the machine can do for opening and cleaning the cotton. Secondly, at every stage, the process has its demands in terms of what is to be expected from the machine at that place.

In other words, a machine in the beginning may have to play a major role of very effective opening, while that in the middle may have to focus more on cleaning, taking the advantage of earlier opening. Again, the blow room has to be versatile, so that it can process different varieties of cotton varying in their trash content and quality. All these, ultimately decide the type of machines chosen, their total number and position of each in the sequence. The schematic diagrams of some few popular blow room lines are given below.

With two sets of mixing bale openers and waste hopper feeders in **Rieter's blow room** line (Fig. 4.52), it is possible to work two different mixings separately.

A large number of bales can be simultaneously opened in each mixing. A sequence of bale openers followed by auto mixer gives very good blending.

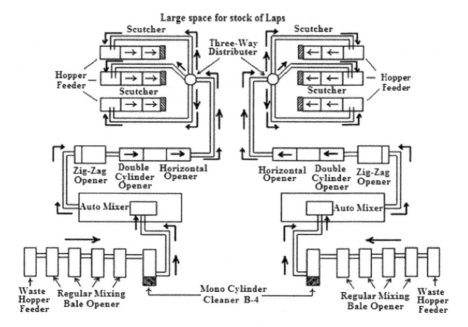

FIGURE 4.52   Rieter's Single Process Blow Room Line[4.6]: A speciality of single processing is that when the lap-slabs are fed at the back, no man handling is required till the lap is formed in the front.

As against mild action of mono-cylinder opener and zig-zag opener, double horizontal cylinder opener can be very useful with cotton having higher trash content. The cotton is finally distributed through a three-way distributor to three Scutcher lines where the laps are formed. The conventional blow room line always needs a sufficiently large space in front of the scutchers to store the laps.

When the 100% man-made fibre laps or blended laps are processed, they require additional separate storage space. However, these laps are wrapped around with polyethylene sheets so as to protect them from any contamination.

In **Platt's blow room** sequence (Fig. 4.53), four mixing blenders and one waste blender are able to give thorough blending. The beaters like step cleaner and porcupine are used to provide very effective cleaning points. In very old blow room line, vertical beater (Crighten opener) was added simply to tackle cotton with much higher trash content. It has been observed that this machine gives harmful action to the fibres and as such can be easily replaced by versatile opening-cleaning machines. The speciality of Platt's line is an inclusion of Air Stream Cleaner to give aero-dynamic cleaning. Finally, the two-way distributor feeds the cotton to two Scutcher lines, each of which include three-bladed beater, Kirschner beater and lap forming unit.

In both the above lines, another important feature is by-passing arrangements which facilitate omission of typical cleaning machines (Vertical Opener or Porcupine Opener) that are likely to give harsher action to finer cotton. Especially,

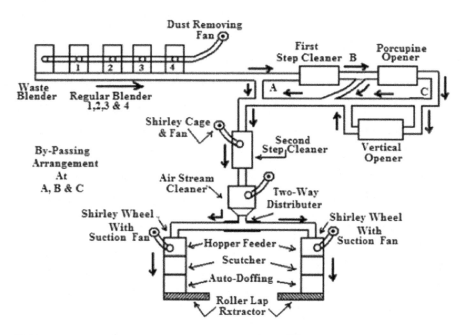

FIGURE 4.53 Platt's Single Process Blow Room Line[1,6]: The speciality of this line is it uses a totally new machine – Air-Stream Cleaner, which is based on aero-dynamic principle of separating lint and trash.

when a mill processes variety of mixings, it is essential to choose the sequence for processing different grades of cotton and with varying trash. Hence some of the machines need to be either included or by-passed.

In conventional blow room lines, it is the number of machines which is varied when processing cotton with different trash content. This improved the versatility and usefulness of the sequence. Platt's suggested some variations to cater to the need of processing material with varying trash content (Table 4.2).

TABLE 4.2

Sequence Used for Cottons with Varying Trash

No.	Low Grade 7–8% Trash	Medium Grade 4–5% Trash	Fine Grade 1–2% Trash
1.	Blenders	Blenders	Blenders
2.	Step Cleaner – 1	Step Cleaner – 1	Step Cleaner – 1
3.	Porcupine	Porcupine	Air Stream Cleaner
4.	Vertical Opener	Step Cleaner – 2	To-way Distributor
5.	Step Cleaner – 2	Air Stream Cleaner	Hopper Feeder
6.	Air Stream Cleaner	Two-way Distributor	Scutcher
7.	Two-way Distributor	Hopper Feeder	–
8.	Hopper Feeder	Scutcher	–
9.	Scutcher	–	–

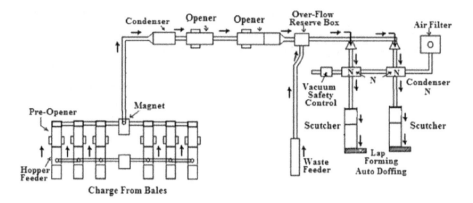

FIGURE 4.54 Another Single Process Blow Room Line,[6]: In this line, Two Openers are used to carry-out cleaning job. Finally, two-way distributor feeds the cotton to two scutchers.

Another Single Process blow room line (Fig. 4.54) consists of six hopper feeders for mixing different lots, with a separate provision for processing soft waste. The most interesting part of this line is that the material from the waste hopper feeder is directly sent to a stage where some preliminary openers are omitted. This way, over-treating of soft waste is very conveniently avoided. The normal lots are supplied to hopper feeders by loftex chargers, which contain heavy-duty aprons.

Thin slabs of cotton from different bale lay-downs are received by the hopper feeder. This ensures that all the bales are uniformly represented. It leads to higher mixing efficiency and improved homogeneity of a blend. On its way to condenser, the material is suitably directed over strong magnets which remove any magnetic metal particles.

An opener can be like a step cleaner with five beaters arranged at an angle of 20°–25° to the horizontal. Though mild in its action, the machine is very versatile and useful in extracting trash. The material is then sent through two-way distribution system to scutcher. Finally, the laps are made and auto-doffed.

4.23 USE OF GRAVITY SEED TRAPS

Often, there are difficulties posed in blow room, when the bale-cotton contains broken seed particles. This is not only peculiar with low-grade varieties but also with higher grade cottons. The seeds usually have fibres attached to them and hence, it becomes difficult to remove them easily.

A wider setting between adjacent grid bars can remove seeds, but this also leads to higher lint loss. When the seeds are not removed in the earlier blow room processing sequence, they are calendered at the final lap making and get crushed. The seed oil is thus allowed to ooze out. This, in a way, contaminates the cotton and leads to many subsequent processing problems like loading on card cylinder and lapping around drafting rollers.

This critically and seriously affects the performance of these machines. The main principle in designing seed traps (Figs. 4.55 & 4.56) is that the trap can be placed in the path of cotton, especially at the place where chute pipes carrying the normal cotton flow bend around at considerable angle (similar to aero-dynamic separation).

The seeds being comparatively much heavy do not follow the sharp bends and thus get deviated from normal curved path. A sliding adjustable door is provided at the bend to control the opening of door to adjust the level of extraction. Sometimes, just around this bend, a magnet (Fig. 4.57) is also suitably positioned and it is able to remove any metallic particle in the cotton.

The separation of metal particles in this manner, earlier in the process, avoids any possibility of damage to any machine parts and, at the same time, prevents any fire hazards owing to possible sparking. The metal particle detection is very important. This detection and separation of metal particles is required to be done much earlier in the blow room sequence, whereas seed traps can be placed at any convenient positions prior to Scutcher.

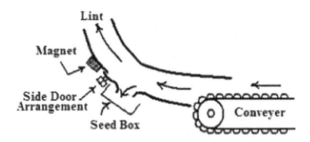

FIGURE 4.55   Seed Traps[6,7]: They allow separation of seeds from main stock. The seeds, if get crushed are likely to contaminate cotton in process.

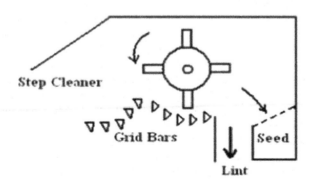

FIGURE 4.56   Seed Traps[6,7]: They are placed much earlier in the blow room sequence of machines. During heavy calendering at scutcher, the seeds, if they persist, get crushed when oil oozes out.

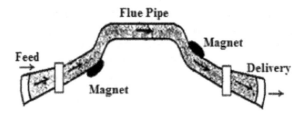

FIGURE 4.57 Metal Extractor[6,7]: During baling and when the bales are later opened in blow room, some metal particles in the form of nails, or broken metal strap pieces are likely to get associated. If they are not separated much earlier in blow room, they may lead to sparks and fire.

4.24 RELATED RESEARCH[3,6]

Opening and cleaning are the two major functions of any blow room. The mechanical or technological short-comings in any machine in the sequence seriously affect the performance of the whole blow room. Apart from the speeds, settings and air control around the beater zone, the sequence of blow room machines, the type of cotton processed and trash content are some other equally important aspects governing these objectives. The choice and the selections of the machines in a sequence decide the type of treatment that the cotton receives and this ultimately affects the trash extraction at blow room. In a sequence, the trash removal efficiency depends upon both the trash in the cotton fed and position of the beater in the sequence. Obviously, when the trash in the material fed is more and if the machine is placed earlier in the sequence, there is always a better trash removal. In short, the cleaning efficiency of the machine is good.

Cleaning Efficiency of machine or process

$$= \frac{\text{Trash in the material fed} - \text{Trash in the material delivered}}{\text{Trash in the material fed}} \times 100$$

1. **Position of the Machines:** Though the cleaning efficiency is expressed as the ratio of trash removed in the process to that fed, the cleaning carried out by a machine placed at different positions in a sequence cannot be expected to be the same. This is because there are far more opportunities for the machines to easily extract the foreign matter and trash when placed earlier in the sequence.

   In one of the trials in the mill, two Step Cleaners were used in the line. These were identical in their construction, speeds and settings. However, the second one used little later in the blow room sequence continuously gave lower cleaning performance. Similarly, in another instance, when the feed order was changed so that the material from blending hopper (initial stage) was first fed to two porcupine openers in tandem and then to Crighten, the performance of porcupines improved, whereas that of Crighten showed a slight deterioration. This clearly indicates the importance of position of the machines in a blow room sequence.

The machines like Axi Flow, which work on aerodynamic principle, also need to be placed at correct position in a blow room sequence. This is because; the cotton has to be sufficiently pre-opened before feeding it to this machine. Therefore, inclusion of this machine without giving due consideration to its proper positioning, results in producing stringy cotton and this is usually experienced with fine and superfine cottons. Studies made in this regard, also revealed that the level of openness of cotton fed to this machine significantly influences its performance. Insufficient opening prior to this machine resulted in cotton remaining for longer time within the machine. This happened especially when it was positioned much earlier (possibly immediately after blenders) in the blow room.

2. **Openness of Cotton:** The level of openness obtained in the conventional blow room sequence is of gross nature. The experiments show that there is a wide disparity in the openness imparted to the cotton. However, it does not appear to have significant effect on the yarn quality in terms of yarn strength, evenness and its performance at ring spinning. In one of the mills, a blow room gave adequate opening with two Crighten opener and three other beaters. The breaks in the final spinning however increased in spite of good opening in the blow room. But the higher breaks were due to over-beating, especially by the Crighten openers.

Therefore, the method of achieving openness is also important. Increasing openness by using a large number of beating points or wrong types of beaters often leads to generating much higher number of neps. However, when cotton is opened by using finer pins or saw tooth rollers, much better cleaning is possible and this helps in significantly reducing card waste. The openness can be measured by the following methods.

1. Average weight of the tuft, when randomly picked. 2. Specific buoyancy of the tufts. 3. Tuft volume. 4. Specific area of the tuft. 5. Resistance to air offered by cotton and 6. Weight of a given volume of cotton.

LITERATURE REFERRED

1. Manual of Cotton Spinning – "Opening & Cleaning" – Vol II, Part II – W. A. Hunter & C. Shringley, The Textile Institute Manchester, Butterworths, 1963.
2. Cotton Spinning by William Taggart.
3. Process Control in Spinning – ATIRA Publication.
4. Rieter Machinery Booklets.
5. Lakshmi-Rieter Pamphlets.
6. Elements of Cotton Spinning – Dr. A. R. Khare, Sai publication.
7. A Practical Guide to Opening & Carding - W. Klein, Textile Institute Manual of Textile Technology.

5 Modern Blow Room Machinery

5.1 CONCEPT

As compared to conventional blow room, the modern blow room looks much more compact with very few machines in the sequence. This shortening of the sequence is not at the cost of any sacrifice in its opening and cleaning action. In fact, it is much improved. The concept of the very first machine – Automatic Bale Opener – has two important advantages: (a) Tremendous reduction in the mass of cotton tufts as they are picked up from bales and (b) A very high potential for mixing cotton-bales representing many lots. This initial opening is able to save at least 3–4 machines of conventional blow room line.

In fact, a correct measure of this initial opening action and the things to follow subsequently can be estimated in terms of tuft size. A tuft size of 2–3 mg is considered to be quite satisfactory in the initial stages of blow room; whereas, when the material is delivered at the end through chute feed, the tufts are reduced to a much smaller size (<0.1 mg).

For this, it is essential that after the initial level of opening, each of the subsequent machines must carry on further to open the cotton tufts. Therefore, the sequence of the machines after bale opening is chosen in such a way that each subsequent machine further reduces the tuft size and significantly increases its surface area (more area). This results into substantial extraction of trash from cotton. In general, a simple principle is adopted – 'the smaller the tufts, more is their surface area exposed for cleaning'. Ultimately, it improves cleaning efficiency appreciably.

5.2 AUTOMATIC BALE FEEDING & OPENING[7]

The opening of cotton tuft in the early part of the blow room line plays an important role. The desired level of blow room performance can be achieved only when the sequence meets this requirement. The important object of blow room, therefore, is to attain a tuft size as small as possible so that the required level of cleaning, without stressing and damaging the fibres, can be easily achieved. Another objective achieved with smaller tuft size is that when the tufts picked up from different bales are very small in the preliminary stages, they have better chance to mix with each other. This greatly improves blending intimacy and also uniformity of the product.

In conventional blow room line, the back feeder was employed for feeding the material from bales. The work basically involved – taking small cotton slabs from each bale and laying them on the creeper lattice of bale opener or blending feeder.

It was necessary to limit and control the weight of the slab that the worker would pick up from each bale at a time. This was because, only then, it was possible to improve the chances of intimate blending. Thus, the whole operation was highly subjective and depended largely on the skills and sincerity of the person-picker. Even with a battery of 4–5 blenders used in conventional set-up, long term blending was not adequately taken care of. Whereas in modern blow room, machines like Uni-Mixer or Multi-Mixer, improve the long term blending only because a very fine opening of bale cotton is carried out earlier by Automatic Bale Opening Machines.

5.2.1 STATIONARY BALES[8,9]

The Spring (or Blending) Grab installation (Fig. 5.1) can process bales in the group of six with different bale heights. The machine operates with spring grabs which can be turned through 90°. The extracting carriage with many grabbing springs is made to move on two rails, along the bale lay-down. At each bale, the carriage stops. The grabbing springs open out and move down to pick up the material from each of the bale surfaces. After a pre-set interval, the grabs close and rise up. They are made to turn in 90° and again open out to deliver the material into weighing container. Again the grabs turn into normal position and the carriage moves to the next bale.

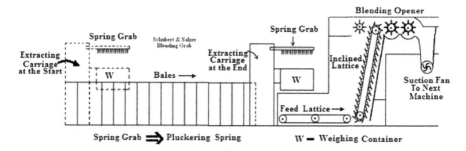

FIGURE 5.1 Blending Grab[4,8]: The fork of the grab, working intermittently, penetrate a little into bales and pluck small tufts of cotton from the bale surface.

Whenever two dissimilar fibres are to be blended, the grab begins its journey in picking-up the fibre tufts from the bales of one component. The process continues till the weight of this component, thus picked up, is as per the blend composition. The carriage then moves to the next component and continues the process with corresponding bales again till the weight of that component is as desired for the blend. Finally, when the weighing container receives the pre-set quantity from each component, the carriage ends its journey, while the material in the container is thrown on to the conveyor lattice of blending opener.

5.2.2 MULTIPLE BALE OPENER: (HERGETH-HOLLINGWORTH)[8]

In this machine, the picking-up of the tufts is continuous. The machine consists of a big hopper and the conveyor placed below (Fig. 5.2). On this conveyor, the bales

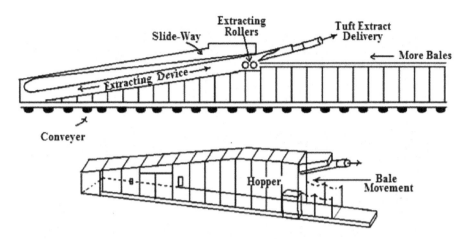

FIGURE 5.2 Multiple Bale Opener[4,8]: Extracting opening rollers works continuously on the bale surface and extract fine cotton tufts.

are laid. The slide-way has extracting device arranged at an angle above the bale surface. About 12–15 bales can be arranged, at a time, on a conveyor.

The extracting device has two rollers that travel at an inclination above the bale surface and extract the material in wedge shape. Depending upon the extracting depth, the material is picked up from the bales and when the bales, in turn, thus get exhausted, the conveyor pushes the new full bales through a small amount from the other side. This establishes the availability of the full bale lay-down. The material picked up by the extracting rollers is then directed to subsequent processing machine.

With the introduction of Blendomat, the requirement of picking very fine tufts from bales is more or less completely met with. The bales are placed directly on the conveyor belt for an initial lay-down. As can be seen (Fig. 5.3), all the bales in the beginning are of the same height. When the detacher unit moves to and fro on the bales, the diagonal, as shown in the figure, is formed. The process of forming this diagonal is very slow and requires large number of cycles and reciprocating motion of detacher unit. Only when a working diagonal of 4–10% (the tapered end is lowered by this margin) is formed, the Blendomat is automatically switched to normal mode. In this mode, it starts working on very small cotton tufts from bales which are laid down. The operation is in one direction and is continuous.

In this mode, depending upon the total number of bales in the mixing, the new fresh bales from untapered end (near the biggest existing bale on the extreme side) are brought into position from time to time. Obviously for this operation, the conveyor belt on which the new bales are put is required to be slowly moved. The new bales are, thus, put on this belt and are slowly brought into the process.

When a mixing is terminated, the conveyor belt stops. No more bales are carried into the process and the detacher unit changes itself from normal mode and returns to horizontal position. This again is done slowly over a large number of travels so that the diagonal which is formed earlier disappears. This horizontal position is timed when the detacher unit is working on the last remnants of the bales. With this,

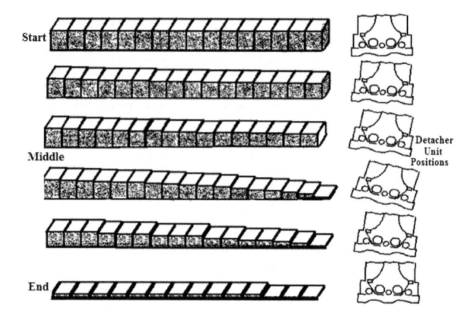

Start

Middle

Detacher
Unit
Positions

End

FIGURE 5.3 Blendomat[4,8]: Initially, the detaching unit steadily assumes inclined positions. During this, a slanting bale surface is prepared. Once a certain inclination is reached, there onwards, the fresh bale feeding from one side and exhaustion of bale from the other side continues till the end of the lot.

when the detacher unit finishes all the bales, the Blendomat becomes ready for the next lot.

5.2.3 BALE FEEDING IN BLENDOMAT[7]

The Blendomat machine is designed for continuous bale feeding. Initially with a full stretch of bales laid down, the bales are added from one end continuously. As shown in Fig. 5.4, bale carriage at the terminal end, or so-called the loading end, takes an opened bale in a forked-lift truck and transports it to the bale lay-down area. With this reloading, the bale reserves can be refilled within a fairly short time. The empty carriage-truck then returns to its original position for carrying the next opened bale to the lay-down area. This results in bringing the newly opened fresh bale into working area. Around this time, the last bale at the other end of the taper is in the position to get exhausted. Further, as the new bale is positioned on a slight slope (towards the already laid-down bales), it does not fall over.

The Blendomat is equipped with safety system (Fig. 5.5). Two parallel light barriers protect the area of operation. If the light barrier is interrupted, all the drives are immediately switched-off and opening rollers are instantly stopped. However, the second area of operation still remains accessible for setting a new and fresh bale lay-down. Similarly, a fire warning and extinguishing device can be integrated in every Blendomat machine. For this, the turret (Fig. 5.5) carries a built-in carbon-dioxide bottle.

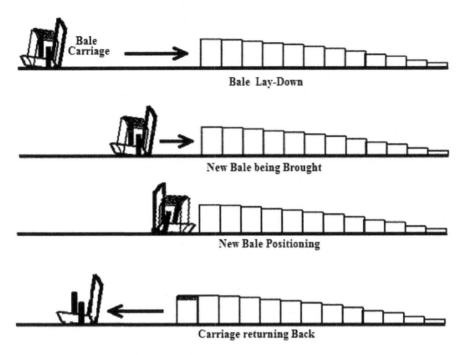

FIGURE 5.4 Bale Feeding[4,8]: As the bales gets exhausted from one side, the carriage-truck moves out and in, to bring a fresh bale. The process continues till the completion of the lot.

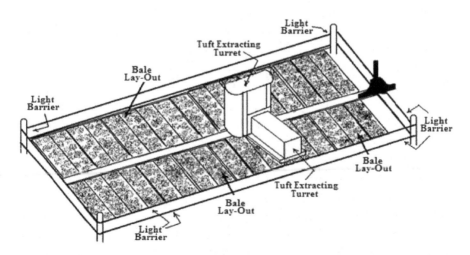

FIGURE 5.5 Bale Lay-out, Tuft Extraction & Light Barrier[4,7]: During working, turret moves to & fro to pluck the cotton tufts from the bale surface. The light barrier, if broken by any human being, immediately stops the moving turret.

5.2.4 DETACHER UNIT

Blendomat is fitted with two opening rollers A and B (Fig. 5.6). Depending upon the direction of the travel, one of the rollers always works on the bales underneath. This can also be seen by observing the direction of the spikes on these rollers. These spikes plunge into the top surface of the bales and pick up tiny tufts from it. The teeth do not directly come against the bale surface, but are made to penetrate through the slits of the grid (C) which are placed just underneath the respective roller spikes. Very small and fine tufts are picked up by the teeth and are subsequently thrown in the suction zone above.

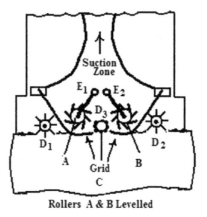

Rollers A & B Levelled

FIGURE 5.6   Detacher Unit[4,7]: The fine curved spikes of opening roller pluck very small cotton tufts from bale surface.

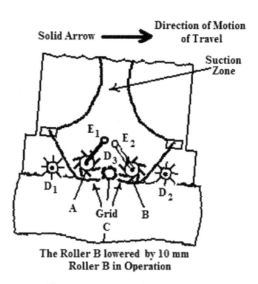

The Roller B lowered by 10 mm
Roller B in Operation

FIGURE 5.7   Detacher Unit[4,7]: The inclination of detacher unit facilitates effecting plucking of cotton tuft from bale surface.

As the detaching unit travels (Fig. 5.7), the roller B is lowered by a small margin of 10 mm. This makes the roller B to strip-off the tufts from the bale surface. It is interesting to note that the spikes of B at the time of contact with the top surface of the bales actually move in the opposite direction of that of travel. During this journey, the other roller (A) is raised by the same margin.

When the detacher unit moves in a particular direction, the leading roller (whether A or B), as explained earlier, is lowered by a certain margin. It immediately starts working on the bale surface. When the direction of the detaching unit is reversed, the positions of A and B are interchanged (Fig. 5.8). This is called "Inversion Mechanism" (Fig. 5.9). The reason for this typical arrangement is very obvious. If both the rollers were perfectly horizontal, the leading roller would have got more chance to work on the bale surface and thus the bale height would proportionately reduce. Therefore, the trailing roller working later on this reduced height would not be able to pick up the same quantity of material from the bale surface. "Inversion", therefore, guarantees that the production in terms of tufts picked from the bale surface is evenly distributed between the two detaching rollers, depending upon the direction of travel.

In addition, there are three supporting rollers (D_1, D_2 and D_3). They slightly press the bales. While doing this, they hold the bale and support it for detaching rollers to act on the bale surface.

For exercising the pressure, D_1 and D_2 are spring loaded. Apart from gripping the bales, they sense the hard and more matted portion on the bale surface. When such hard portions are sensed, the spring gets compressed and this is relayed to overload signal, which stops the machine. This avoids any possible damage to the detaching rollers A and B. The levers E_1 and E_2 are connected to inversion mechanism.

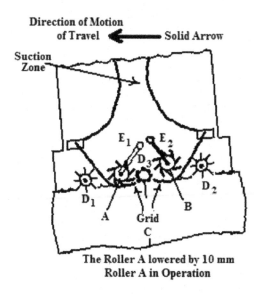

The Roller A lowered by 10 mm
Roller A in Operation

FIGURE 5.8 Detacher Unit Moving[4,7]: The inversion mechanism brings front opening roller in operation; whereas the back roller is lifted-up.

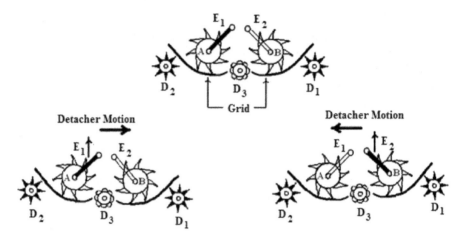

FIGURE 5.9 Inversion[4,7]: The raising of back opening roller or lowering of leading opening roller is only marginal, thus allowing picking-up of tiny tufts.

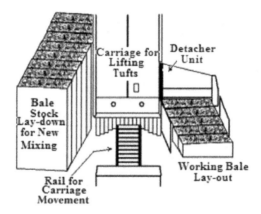

FIGURE 5.10 Detachment of Tufts[4,7]: The present & new bale lay-out are placed with a small distance in between. As the toothed disc (opening roller) plucks very tiny tufts, they are led through.

As shown (Fig. 5.10 & 5.11), the bales can be arranged on either side of detacher unit. The unit moves from bale to bale and picks up tiny cotton tufts from the top surface of the bales.

On the other side, a new bale lay-out can be arranged for the next lot, so as to avoid any time delay. If the lay-out of the bales is on both sides, the detacher unit, after reaching the end on one side is made to turn completely in 180°. It then starts picking-up material from bales laid on the other side. The detaching head with two opening rollers requires approximately 3500 m³/min of suction air. The grid, through which the spiked detaching rollers work, prevents large fibre tufts from being loosened at a time. The work-off (picking the tufts from bales) is carried out in both directions of carriage movement.

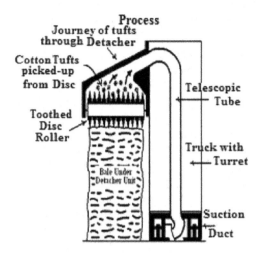

FIGURE 5.11 Detachment of Tufts[4,7]: The present & new bale lay-out are placed with a small distance in between. As the toothed disc (opening roller) plucks very tiny tufts, they are led through.

A special sensor system automatically registers the bale height as well as the beginning and end of the bale lay-down. With different types of Blendomat machines, the production capacity varies from 600 to 1500 kg/hour. As shown in the graph (Fig. 5.12),

1. 1200 mm width, 38 toothed Disc and 1200 r.p.m.
2. 1600 mm width, 50 toothed Disc and 1200 r.p.m.

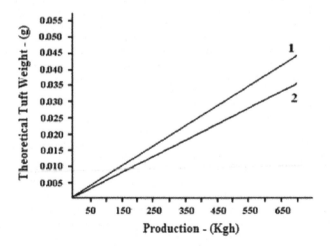

FIGURE 5.12 Theoretical Tuft Weight[4,8]: With increase in the production rate, the opening rollers get insufficient time to work cotton tufts to reduce their size. However, with higher speed and teeth on the disc, this state (from 1 to 2) can be improved.

Very fine tufts (Fig. 5.10) are picked up by the toothed disc. However, the tufts size is governed. on the basis of production rate, So also the opening action is governed by the working width, the toothed disc, its teeth and speed.

The Blendomat can be programmed for working on one side or both the sides. Similarly, it can work in both running directions or in only one direction. The common travel speed of detacher unit is 10 m/min. However, it can be varied between 5 and 15 m/min.

If the Blendomat works in either direction, production rates can be increased by about 25% or alternately, its vertical stroke increment can be approximately reduced so as to produce smaller tufts. Depending upon the blend composition, the bales can be laid down. For example, for 50:50 blend, the bales belonging to each component can be laid alternately. With different bale heights for different components, the bales of each component are grouped and laid together. When all the bales from each group are to be used simultaneously, the bales with more height are worked-off faster. In this case, the stroke increment is proportionately higher.

Working on similar principles, Rieters have come out with A-11 Uni-floc Automatic Bale Opener. Here too, the bales are opened so as to produce very small micro-tufts. Four different assortments can be simultaneously processed with production rates up to 1400 kg/hour. Whereas the take-off width from 1700 mm to 2300 mm is selectable, the total bale lay-down length can be extended up to 47 m.

5.3 SOME OTHER MODERN MACHINES

Though the blow room contributes only 5–10% of the total production cost of spinning department, the loss of raw material in the form of lint which appears in varying proportions in the blow room droppings is a factor to reckon. This is because the blow room cleans the cotton by extracting trash when some accompanying lint is bound to be lost. In fact, the unintentional fibre loss at any stage has significant influence on the cost of the yarn. Equally important is the stress on fibres during blow room processing. In this respect, many modern machines have been designed to extract the trash without putting much stress on the fibres.

5.3.1 RIETER'S UNI-CLEAN[7]

The machine works on similar principles as that of Axi Flow. The striking difference is that whereas there are two beaters in Axi Flow, Uni-Clean has only one beater. As can be seen (Fig. 5.13) with outer covers opened) the cotton passes through the machine across the length of the beater. This means that the rotation of the beater does not support, in any way, the movement of the cotton through the machine.

The cotton enters the beater chamber through an in-let A in free condition. Immediately, the beater spikes (or pins), which are typically shaped, start acting on cotton. The spikes prevent any possible influence of cotton rushing through the machine. With comparatively tiny spikes, cotton tufts are thoroughly opened without causing any harsh action. The actual cleaning takes place when the tufts are

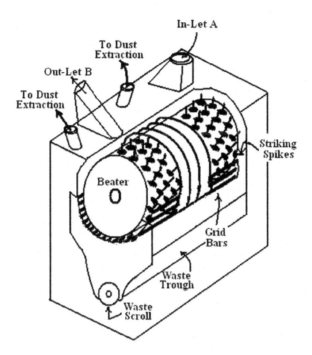

FIGURE 5.13 Rieter's Uni-Cleaner[7,8]: A wonderful cleaner with a provision of fine striking spikes. The beater spikes strike the cotton in the free flow. It is very versatile and yet very effective cleaner.

struck on the grid bars which are suitably placed underneath the lower section of the beaters.

In traditional beaters, the beater blades while striking the cotton against the grid bars also push the cotton ahead. Thus, their rotational direction helps in assisting the material to pass through and out of beater zone. Whereas in Uni-Clean, the cotton moves across the beater length and, during its journey, gets repeatedly struck by the beater spikes and simultaneously is thrown against the grid bars. In this case, not only good amount of cleaning is possible, but the force of cleaning is also not very harsh. Therefore, a gentle and yet effective opening followed by efficient cleaning makes this machine acceptable in any blow room sequence.

A special dedusting element is incorporated within the machine to extract impurities, especially of heavy nature (like sand). Finally the waste collected underneath the grid bars is conveyed to a transport system by a scroll roll, which takes away the droppings. The waste is either sucked intermittently or continuously. The machine is operated under 'Vario-set Cleaning Field'. Thus, with only two settings, it is possible to change the severity of the action (gentle to intensive) within the beater chamber. The 'vario-set technique' translates the desired operational characteristics into the actual values of settings. The two values in Vario-set are related to (1) speed of the beater which decides the intensity and (2) the relative amount of droppings, which is influenced by the grid bar angle. The novelty of this operation is that, even when the machine is running with the material, the waste extraction level can be altered and controlled. Whereas, in conventional beaters, the machine is required to be stopped for these settings, the operation with 'vario-set' becomes quick, easy and without any time loss. The material is finally delivered through an out-let B, which leads the cotton through a suction system to the next machine in sequence.

The working dimensions of the machine are 1.6 m (width) and 0.75 m (Cylinder dia.). The cylinder speed can be varied from 480 r.p.m. to 800 r.p.m. and the production up to 1200 kg/hr is possible.

5.3.2 CLEANOMAT SYSTEM[7]

The optimum efforts to clean the cotton has become the need for the production of quality yarns. The Cleanomat system fulfils this demand of maximizing cleaning efficiency and yet offering most gentle fibre treatment. The machine is equipped with different rollers (cleaners) on which different types of cleaning contrivances like spikes, needles or saw teeth can be mounted. Depending upon, therefore, the use of these elements, the class of opening can be selected for processing. Alternately. depending upon the class of the cotton to be processed, the cleaners with a suitable set of cleaning elements can be selected.

As shown (Figure 5.14 [a–d]), there are four positions of deflector blades. With the plate closest to the roller, both the lint and trash are also kept close to the roller and their separation is less effective. The progressive widening of the deflector plate

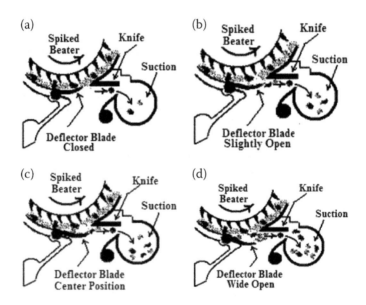

FIGURE 5.14 The typical various positions of Deflector plate & Trash Extraction: The deflector plate having sharp edge is responsible for separating lint and trash. This happens when cotton is well opened by spiked beater and presented to deflector plate. By altering its positions, the proportion of lint and trash in the trash box can be varied: (a) fully closed; (b) slightly Open; (c) centre position; (d) wide open.

allows trash, which is in general heavier than lint, to move away from the roller surface. This is owing to difference in gravity between lint and trash. The higher centrifugal force leads the trash to go away from the roller surface. This bifurcation is almost completed at the following mote knife region. At this point, the trash particles, thus separated earlier, come directly under the influence of suction suitably generated (inside suction hood) just below and around the mote knife. The Cleanomat system due to these peculiarities is most suitable for processing sticky cottons. This is because, unlike the traditional cleaning devices with grid bars where the sticky dirt, even when freed, settles and accumulates on grid bars, in Cleanomat, it is immediately sucked away.

In fact, in none of the other cleaning devices the suction provided to take away the liberated impurities is so closely placed. The Cleanomat system with the single cleaner roller, the feed arrangement, the position of deflector plate, mote knife and suction hood are all shown in Fig. 5.15.

The machine has a long feed lattice carrying material in loose form. The pressure rollers merely assist the movement of material at a regular pace. A pair of feed rollers feed the material to the cleaner roller. The needles or saw teeth of the cleaner immediately start acting on the material flock.

It is here that the mote knives, coupled with suction hood, play an important role. Whereas the beater needles open out the flock, the knives greatly assist in a thorough cleaning action. The opening power of this type of cleaner is exactly tuned to the feeding requirement of the last cleaner (like Kirschner in conventional) in modern blow room sequence.

A similar cleaner CVT is equipped with the spikes for comparatively heavier work (Fig. 5.17 [a, b]), where the coarser and medium trash particles are very effectively separated out. As against this, the needles with their fine points (Fig. 5.18 [a, b]) used in CNT cleaners, lead to comparatively finer work. The principle used here is extended in different versions where number of cleaners is increased. Thus, in CVT/CNT-3 there are three cleaners. Here the first cleaner is the

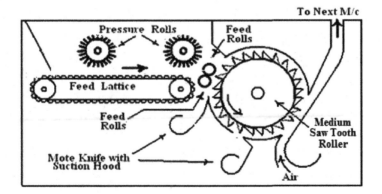

FIGURE 5.15 Feed to Cleanomat System[4,7]: The feed is invariably in a fleece form when the spiked/saw-tooth beater is made to strike and open the fleece to very tiny tufts. This results in very effective cleaning.

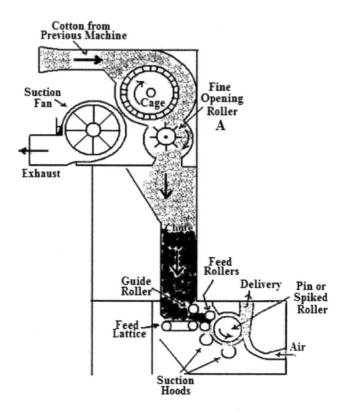

FIGURE 5.16 Other Feeding Method to Cleanomat: Alternately, Cleanomat can be fed with controlled feeding system. This improves the opening & cleaning power of subsequently placed Cleanomat.

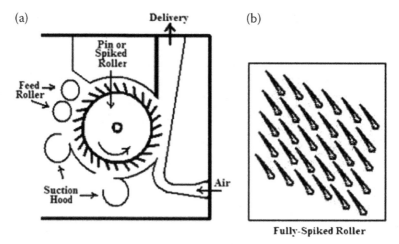

FIGURE 5.17 CVT or CNT Cleners[4,7]: If the machine is used a little earlier in blow room, the beater is covered with coarser spikes. The tuft size in this case is comparatively bigger. (a) spike covering; (b) spikes.

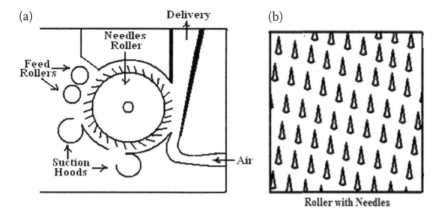

FIGURE 5.18 CVT or CNT Cleaners[4,7]: They almost look alike. With some opening already received by the cotton, the finer needles do a superior opening work to bring out finer cotton tufts. (a) needle covering; (b) finer needles.

same as in CVT/CNT-1. However, the second and third rollers in CVT-3 have coarser and finer saw tooth wires, respectively (Fig. 5.19). As shown in (Fig. 5.16), cleanomat can also be fed with the help of high-capacity condenser.

Correspondingly, in the second and third rollers in CNT-3 (Fig. 5.19) the medium and fine saw tooth wires are used. It can, therefore, be seen that in both these types, the opening power of the cleaners is progressively improved and increased. The material after leaving the final cleaner directly enters the vertical delivery chute and is finally fed to the Card. This is materialized by placing a suction fan at the end of this delivery chute.

It may be seen [arrows for air in-let shown in Fig. 5.17 a, 5.18 a] that the air for this suction is not received through the cleaning zone of either single cleaner or three cleaners (Fig. 5.19). Separate passage for the air is provided so that the air being sucked directly enters the delivery chute without affecting the cleaning performance of any cleaner.

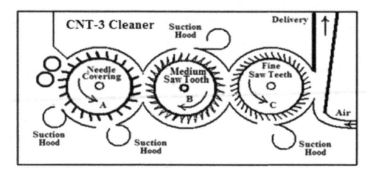

FIGURE 5.19 CVT and CNT[4,7]: In both, when the number of cleaner are more than one (in Fig. 5.19 there are three cleaners) the spikes or needles of the cleaners progressively become finer. This results into steadily increasing opening and cleaning action.

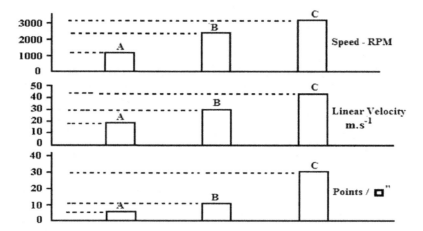

FIGURE 5.20 Comparative Particulars of CVT-3 Cleaners[4,7]: It can be seen that the speeds, linear velocity and the point density progressively increases from first to third cleaner. This results into steadily increasing opening action of the cleaners.

However, while passing by, the air gives a brushing-off effect to the saw teeth or spikes/needles of last cleaner, thus carrying the opened material into delivery chute without allowing it to clog on the spikes or saw teeth. Fig. 5.20 shows a progressive increase in three cleaners of CVT/CNT-3 in respect of their speeds, linear velocity and points density. This reduces the harshness in the cleaner action but retains the effective opening and cleaning power.

CVT-4 (Fig. 5.21) is built-up on the same lines and has four cleaner-rollers arranged in a sequence. The first cleaner is clothed with spikes. The second, the third and the fourth cleaners are equipped with coarser, medium and finer saw teeth, respectively.

The delivery from the final cleaner is arranged with the help of inverted 'U' tube so as to facilitate entry of air, which again helps in sweeping the material from final

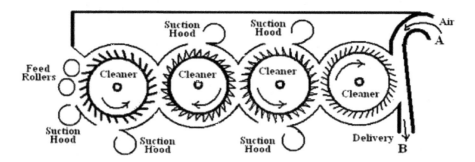

FIGURE 5.21 CVT-4 Cleaner[4,7]: With each additional cleaning roller, the power of CVT/CNT cleaner in opening and cleaning the tufts is improved. The needles of the fourth cleaner are very fine and this results into finest tufts.

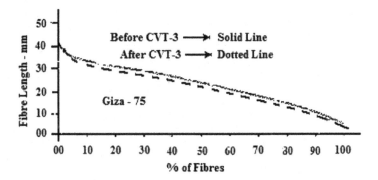

FIGURE 5.22 Fibre Damage[4,7]: Almost similar Baer Sorter diagram conclusively proves that there is not much of fibre breakage when CVT (or even CNT) is used in the process. This is because the spikes/needles or even saw tooth points do not cause serious harm to cotton tufts (no fibre damage). **Baer Sorter Diagram**.

TABLE 5.1
Types of Cleanomat & Production Capacity

Machine Type	Production – kg/hr	Machine Type	Production – kg/hr
CVT-1	525–770	CNT-1	400–510
CVT-3	350–440	CNT-3	350–450
CVT-4	250–440	–	–

---

roller and carrying it upwards to its next destination. There are two different types of feeding systems (as shown in Figs. 5.15 & 5.16).

The fine opening roller A (Fig. 5.16) diverts the material into a feed trunk, which ultimately deposits the same on a feed lattice. As a variation to this (Fig. 5.15), a normal Bale Opener is used to deliver the cotton on to the traditional lattice which feeds the material to the Cleanomat system.

5.3.2.1 Production capacity of Cleanomat Machines

In the initial stages of the introduction of Cleanomat machines, especially with CVT-3 and CVT-4, there was apprehension about the possible damage to the fibres, especially when so many beaters, one after the other were used. It needs to be mentioned here that the fineness of the striking elements (spikes or pins) is progressively increased in these machines. Therefore, what is stressed in their action is a gradual and a progressive opening of the cotton tufts. Obviously, what follows in these machines, as a result of such effective opening action, is the cleaning, which is completed by the suction hoods appropriately and suitably positioned around the openers. From the Fig. 5.22, it is evident that the fibre length diagram before and after CVT-3 treatment is almost the same, confirming that there is no fibre breakage or damage caused by Cleanomat system. In fact, owing to their great opening power, the Cleanomat system can also be deployed in recycling of waste.

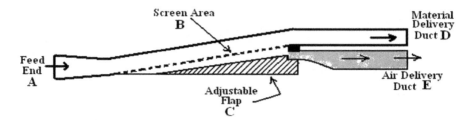

FIGURE 5.23   Air Separators[7,8]: They enable easy separation of lint & trash on the basis of their differences in buoyancy. As no power is needed to operate them, they are very economical trash extracting devices. **Air Separator**.

5.3.3   SEPARATORS[7]

The modern blow room line usually contains some typical cleaning units in which a very high degree of opening along with very effective cleaning is achieved. Especially when the cotton is transported to the next machine in such a good opened condition, it is possible that very fine impurities which escape cleaning action in the preceding machine are still in the separated condition.

Therefore, before these separated fine impurities get a chance to recombine (at next machine) with cotton, they need to be conveniently removed during the transit. It is here that the Air Separators play an important role. While providing a partial suction to help the transportation, they separate the lint and impurities. As shown in the Fig. 5.23, the material enters the separator at a feed end A. The leading pipe on under side has a screen area (B). The lint accompanied by fine impurities passes over this region. An adjustable flap C (shaded portion) is provided to adjust the gap for the air leading to dust-laden air-delivery duct (E).

The material (lint), however, follows the normal path over the screen and is delivered through the normal delivery duct (D). The absolute levels of the two air currents – one for removing the dust-laden air from the cotton coming from preceding machine and the second for feeding the material to the following subsequent machine are carefully controlled. There is truly no maintenance required for the installation and an additional benefit is the dedusting of cotton.

The functioning of yet another Air Separator is shown in Fig. 5.24. The material is fed into the device through in-let (1). It can be easily seen that the side screen area (2) along the vertical passage helps in dust extraction at A. The dust is collected and passed through the dust extraction out-let (4). The cotton is finally delivered to the next machine through the out-let (3).

5.3.3.1   Separomat ASTA[7]

M/s Trumac have come-out with yet another simple device to separate the heavy impurities. As shown in Fig. 5.25, the material, usually from a very fine opener, is introduced at B. This is done by a suction created by a fan placed subsequent to this device. The material entering at B, is expected to be a mixture of fine tufts and loosened-out heavy trash particles. The Trash being heavy, it follows the diversion path at C.

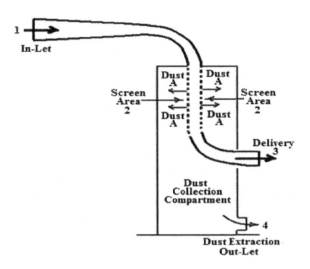

FIGURE 5.24 Air Separators[7,8]: They enable easy separation of lint & trash on the basis of their differences in buoyancy. As no power is needed to operate them, they are very economical trash extracting devices. **Air Separator**.

The bend at D almost completes the separation of heavier trash and lint. The lint, however, follows the bend and moves upward, along with the normal lint, towards an exit E.

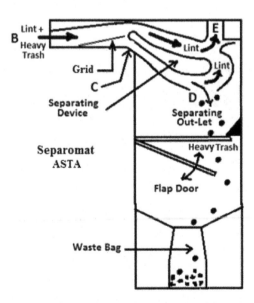

FIGURE 5.25 Separomat ASTA[7]: It also works on the principle of separating lint & trash on the basis of differences in their buoyancy. The trash being heavy, drops down in the waste bag placed below.

The heavy trash is let through separating out-let and gathered over the flap door, which intermittently moves around to drop the trash. When the waste bags below are full, They can be emptied at intervals. As claimed by the manufacturers the device is totally power-free and requires no maintenance. The flow of the material through the device can be as high as 800 kg/h and even at such a high rate of output, the device can perform quite satisfactorily to separate the heavier trash.

5.3.4 Dedusting Machine[7]

It has already been mentioned that the most important pre-requirement for effective separation of normal dust, and micro-dust is a very good opening action preceding the device. There are typical cleaners like CVT and CNT in modern blow room which, with their fine needles or saw teeth, lead to a very thorough opening of cotton tufts. The positioning of De-dusting machine is, therefore, done immediately after such efficient cleaners. The functioning of Dustex machine is shown in Fig. 5.26.

The material enters Dustex through a feed duct (A). The suction fan (B) helps in this action. The material is made to travel vertically up by a guided framing (A') forming a pipe and enters a chamber (H) through distribution flap (C) In the chamber, the material is thrown over a perforated screen plate (D). This plate is typically shaped and when the cotton falls over perforated screen, it slides down over the perforated area due to gravity.

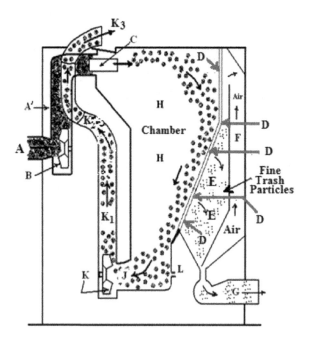

FIGURE 5.26 De-Dusting Efficiency[7]: Basically it depends upon the strength of the suction. The higher the suction, the greater is the dedusting efficiency.

The suction is provided through the perforations. Thus, the small cotton tufts and released fine dust are exposed to this suction. A dust suction fan is provided to act through the region (F). The vertical channel called 'waste suction channel' (E) is placed just behind the perforated screen area, and it thus collects the dust and air mixture. Whereas the heavy waste falls down due to gravity, the air is diverted vertically upwards and enters another dust suction zone (F).

The separation of these two compartments helps in avoiding any undue turbulence. Finally, both the fine dust and comparatively coarser waste are removed through waste duct (G) situated at the bottom. The cotton falling down in chamber (H) is collected by a suction funnel (J) and is passed vertically up through a pipe. The suction fan (K) is provided to help the material to be forced up and led through another pipe (K_1-K_2-K_3). A separate air in-let (L) provides the air for this journey of cotton. The provision of this air in-let prohibits any interference of sucked air through the suction funnel (J) and thus avoids any possible disturbance during dust passing over the perforated screen (D).

The integration of dedusting unit in a modern blow room line has gained a lot of importance, especially in the mills having Open End spinning machines. It is already known that the greatest source of trouble when working with O.E. machines is the fine micro-dust. A very effective removal of this kind of waste at the blow room itself improves the efficiency of O.E. spinning machines. Even in traditional ring spinning, there has been marked reduction in the yarn imperfections and breaks during spinning.

The speed with which the material is made to pass through De-Dusting machine has profound influence on the efficiency with which the process is carried out. With the higher production rate, therefore, the 'de-dusting ratio' rapidly falls down.

Thus, the production rate of this machine needs to be precisely controlled to obtain the best de-dusting efficiency. In Fig. 5.27, the three curves represent the parabolic relations of de-dusting efficiency w.r.t. the production rate. The suction employed in the machine is related to the impurities sucked. Obviously, therefore, the curve for the higher suction (4800 m³/h) shows the higher de-dusting efficiency levels for corresponding production rates.

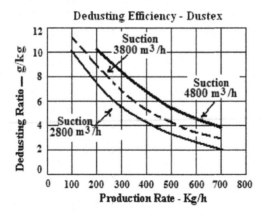

FIGURE 5.27 De-Dusting Efficiency[7]: Basically, it depends upon the strength of the suction. The higher the suction, the greater is the dedusting efficiency.

5.3.5 BLENDING AT BLOW ROOM

The various properties of the ultimate yarn greatly depend upon how intimately the fibres from the different bales get uniformly distributed. It is known that in a modern blow room, it is possible to open out many bales and simultaneously carry-out mixing of cotton. It may be recalled that even with a single component (cotton), there is substantial variation from bale to bale or even within a bale, owing to agricultural practices, region of growth or even climatic condition. These are responsible to vary the fibre properties of cotton grown.

The modern blow room sequence also facilitates mixing of dissimilar fibres (e.g. cotton/polyester). The uniform distribution of either a single fibre-component (cotton from bales) or multiple fibres is possible when automatic bale openers is used (tuft size less than 1 mg.). This is the first stage to the successful mixing of fibre-component/s.

In addition to this, the modern machines like 'Multi-Mixers' used in the line, almost complete the process of intimate blending. It is needless to state here that when such a care is taken much earlier in the manufacture of yarn, itself, not only is the yarn more uniform in its characteristics but also in the fabric processing later, the defects like patches, spots or specks are almost absent. Modern blow room blending thus ensures remarkable homogeneity of a mix.

5.3.5.1 Blending Machinery Equipment[7]

In blow room, when dissimilar fibres are blended, the gravimetric blending becomes necessary to ensure that the component fibres are correctly represented by their weight and not by their volume. This is because of the difference in their densities. As shown (Fig. 5.28), the layers of the material as laid on the feed lattice of Hopper Opener are picked up by the inclined lattice. They are then subjected to evener roller action. The resulting finer tufts pass ahead and are put into weighing pan by a pair of feed rollers.

The arrangement is made so that after the required quantity of material is put into pan, the feed stops. On receiving the signal, the pan opens out and pours the material on to another lattice or belt, which feeds it to a very fast-moving opener. The speed and the fine pin-covering on the roller give a very intensive opening action and simultaneously help in blending the material intimately. The delivery of the machine is through an exit chute, which draws the material up under the influence of a common suction device (a cage or a suction fan).

In a tuft blending, depending upon the number of blending components, the separate weighing pans with Bale Openers are provided. The weighing is carried by the precision electronic scale-pans (Fig. 5.29) with an accuracy of one gramme and, thus, even a smallest blending component can be dosed very precisely. It is possible to get the production rates even up to 1000 kg/h. In addition, there is provision for 'Lot Protocol'. The weighings are carried out through computer, which continuously compares the targeted and actual weighings.

With the computer, 'recipe-memorizing' is also possible and it helps in quick lot changes. The weighing-pan scales are also computerised. When the scales are empty and ready to be filled, the filling rate is high. As they get almost full, the rate

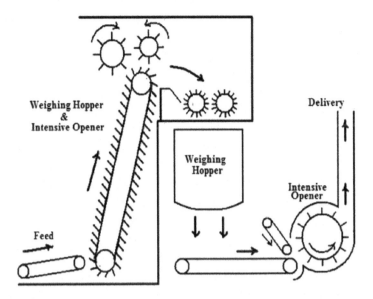

FIGURE 5.28 Weighing Hopper & Intensive Opener[4,7]: The weighing of the different components before blending brings more accuracy in blend percentage. The Intensive opener helps in opening the blend mixture thoroughly to improve blend intimacy.

is automatically slowed down. When the required quantity of the material, as predetermined by the computer is filled, the flap door (Fig. 5.29) allowing the material into weighing pan scale is shut-off and the feed stops. The figure (Fig. 5.30) shows the difference between a draw frame blending and tuft blending. It is claimed that by going for tuft blending at blow room stage, it is possible to at least eliminate one blending-draw frame passage and still get the same intimacy. It can be seen that the tuft blending uniformly distributes and spreads the component fibres and thus leads to more homogeneous and intimate blend.

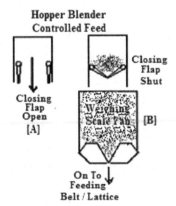

FIGURE 5.29 Owing to precise weighing[7], the blend components are constantly & correctly represented throughout the blend. This reducing patches in the yarn.

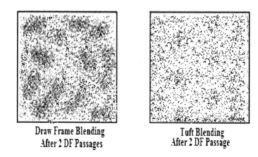

Draw Frame Blending
After 2 DF Passages

Tuft Blending
After 2 DF Passage

FIGURE 5.30 The merging of the two components with tuft blending is so intimate that it leads to great reduction in yarn shade difference.

5.3.5.2 High Volume Mixer[7]

Lot to lot variation in man-made fibre bales is very common. If a small quantity is taken for processing at a time, it can lead to variations in the batches of yarns prepared from these lots.

The High Volume Mixer facilitates blending of a sufficiently large quantity of man-made fibre bale-material, differing in batches. This results into a better homogeneous blending. The material is received through distribution flap at the end of which is a vibrating flap. This distribution flap throws the material over the perforated screen. The fibre dust, if any, is thus removed.

The 'Light Barriers (Fig. 5.31) situated suitably in the hopper automatically control the level of material in it. Therefore, a constant quantity of material within the hopper is maintained. Underneath the layers of the material in the hopper, a

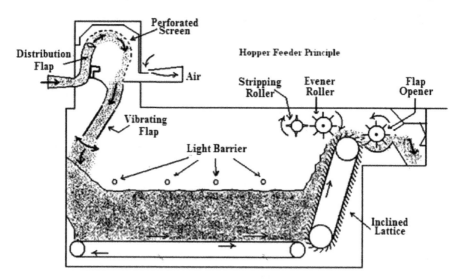

FIGURE 5.31 High Volume Mixer[7]: The hopper accommodates comparatively large volume of blending components, which are laid in the form of stacks. This improves blend intimacy.

slow-moving belt pushes the material towards inclined spiked lattice. The spikes on this lattice are very fine. Ultimately, these spikes pick up the tiny tufts across the whole cross-section of the laid material within the hopper. Ultimately, this brings about a very good homogenous blending.

5.3.5.3 Trutzschler's Multi-Mixer Principle[7]

It is customary in any modern blow room that the blending of components should be carried-out before the action of intensive opening of the mixed components. One such machine is shown in Fig. 5.32. The feed usually consists of mixed material from the different lots. The mixer itself contains several vertical chutes with a flap door above each of them. The height of the material in the chute is controlled by the 'light barrier' and thus there is always a constant level of the material inside each chute. Each chute, in turn, opens out below to an opening system consisting of two guiding rollers and an opener below.

Finally, the opened-out mixture of the lots is sucked through a delivery pipe. In another typical installation, the opener below lays the material on a moving belt situated below. This belt, in turn, feeds the material to a pair of inclined lattices.

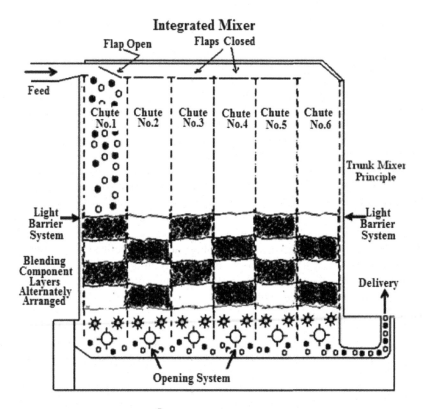

FIGURE 5.32 Integrated Mixer[7]: The blend mixture in each chute is maintained at a constant height. This regulates the amount of mixture laid on the lattice. Each of the opening system further reduces the tuft size and improves blend intimacy.

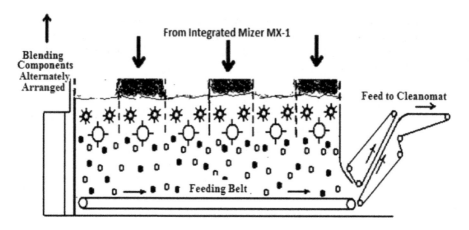

FIGURE 5.33 Feed to Cleanomat[7]: The spikes of the Inclined lattice picked up very tiny tuftes, which are fed to Cleanomat system placed subsequently. This improves the cleaning efficiency of Cleanomat.

Ultimately, the material carried by the inclined lattices is again arranged horizontally and fed to Cleanomat system (Fig. 5.33).

5.3.5.4   Rieter's 3-Point Mixing Principle[7]

The Rieter's B-72/B-76 Uni-Mixers contain eight vertical trunks in which a constant height of fine-tuft material (cotton-flocks) is maintained. After automatic bale opener, Rieter usually has a Uni-Cleaner (B-12 or B-17), which substantially reduces the size of the tufts. The supply of the material into 8 different trunks of Uni-Mixer from Uni-Cleaner comes in the form of different layers representing different lots.

The trunks thus have layers of the material-lots in a stack mixing form. As seen (Fig. 5.34), the material from each trunk at the bottom turns around at 90^0 This leads to staggering of the material in the lots leading to 'First Point' of blending (A). The underneath aprons carry the eight different layers from eight trunks ahead, when the sandwiched layers meet the combing action of inclined spike lattice. The lattice has very fine teeth which pick up small tiny tufts across the eight layer.

This is 'Second Point' of blending (B). The tiny tufts carried-up further by the spike of inclined lattice meet the action of evener-like roller assembly where it is the 'Third Point' of blending (C).

Thus, at three succeeding stages, there is a chance for perfect homogenous blending. As for the quality and economics of the yarn, this becomes a precondition. Basically, a superior homogeneity is one of the characteristics of modern blow room line.

A thorough blending has been possible due to the introduction of Automatic Bale Opener, where the take-off material from Automatic Bale Opener itself is in the form of very fine flocks. In short, uniform blending in blow room ensures

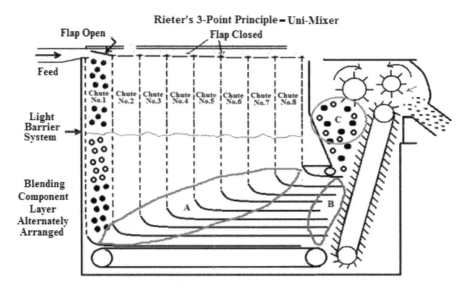

FIGURE 5.34 Rieter's Uni-Mixer[7]: The number of chutes are increased to eight. This is expected to further improve blending intimacy. Here too, the fine tufts are picked up by Inclined Lattice to continue the opening action.

the constant yarn properties over a much longer period and assures a fair price to the yarn, thus manufactured.

5.3.5.5 Card Pre-Opening Rollers[7]

In modern blow room installations, where the cards are usually fed directly through the chute, it becomes very important to see that the tufts gathered in the chute are of the smallest possible size. This reduces the working load on the card and improves its function of individualisation. In such cases, the use of pre-cleaners becomes

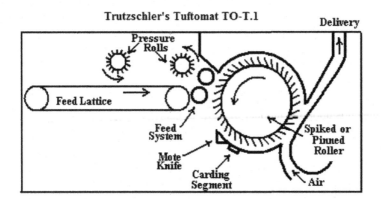

FIGURE 5.35 Tuftomat Opener[7]: The function of pre-openers before carding is to reduce the cotton tufts to the finest size so that it becomes easy to satisfactorily achieve fibre individualisation stage at card.

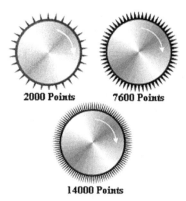

2000 Points   7600 Points

14000 Points

FIGURE 5.36 Type of Pinned Roller[7]: The rollers with both varying point density and fineness of the needles are available. The density and fineness is progressively increased to intensify opening of the cotton tufts.

necessary. The Tuftomat Opener itself (Fig. 5.35) functions on similar lines as that of CVT Cleaner and depending upon the need, there could more than one opener used in the line. The spikes or the pins on the succeeding openers, in such case, are progressively finer in size and dimensions.

One of the special features in these pre-openers is the use of carding segments, almost similar to those used in carding. They not only help in giving extra opening to the tufts but also help in achieving more cleaning. The opening power of the tuftomat type of openers can be varied by using the openers with varying pin densities (Fig. 5.36). The appropriate choice decides the level of openness of the tufts finally led into the chute that feeds the card.

It may be mentioned here that in a typical modern blow room sequence, three pre-openers may be used as the last three machines, before chute feeding device. Each pre-cleaner has pins progressively increasing in their fineness and the pin density. Thus, when ultimately the material is fed into carding chute, the openness of the tuft is completely assured. The graph (Fig. 5.37) reveals the performance of the four types of pre-openers where-in, it can be seen that the production rates through these machines do affect the degree of opening of the tufts. At higher production rate, the degree of opening the tufts is gradually reduced. Thus, it is necessary to restrict the production rate of these machines to a level just needed to meet the card production rate. However, it may be noted that the rate of production itself depends upon the number of cards in chute feeding systems. Even then, when more than one pre-cleaner is used in tandem, it is still possible to increase the production rate appreciably so as to meet the carding requirements.

5.3.5.6 Pre-Cleaner – CL-P (Trutzschler)[7]

It is another type of opening and cleaning machine but slightly different in construction. It is suited for cleaning very heavily contaminated raw material. It has double roller with coarsely spiked covering and as such, complements the cleanomat range of machines.

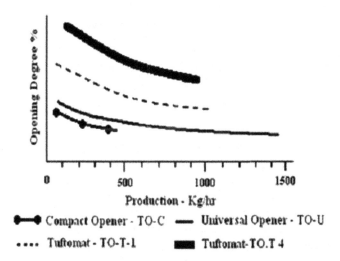

FIGURE 5.37 Effect of Production Rate on Degree of Opening[7]: The higher rate of production gives less time for the opening action of the spike/pin roller, thus reducing cleaning efficiency of the machine.

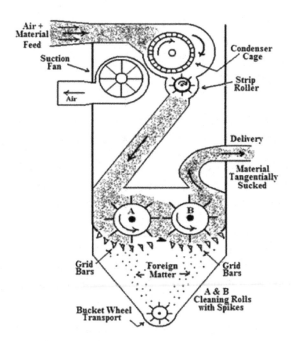

FIGURE 5.38 CL-P Cleaner[7]: A typical Opener/Cleaner where gravity is used to feed the material to the cleaners equipped with fine blades. The adjustable grid bars placed below the cleaners help in further improving the cleaning efficiency of the machine.

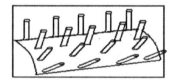

FIGURE 5.39 CL-P Cleaner Spikes[7]: Both the rollers (A & B) have very powerful spikes and are very effective in reducing the cotton mass.

Being slightly coarser in its action, this machine is placed in the early part of modern blow room line, before the action of fine openers. The material is sucked-in by a cage (Fig. 5.38) with the help of a suction fan positioned close to and linked-up with cage. The material released by the cage is received by a strip or a bucket roller.

The function of this roller is merely to see that all the material on the cage surface is released by its spikes into a chute below. A and B are the two rollers which, with their typically shaped blades (Fig. 5.39), work on the material over the grid bar sections and help in opening it. The coarser impurities are released and they make their way through the gaps between the grid bars.

The gathered trash or foreign matter below is transported out of the machine with the help of bucket wheel system and is led into suction duct. The bucket wheel-lock helps in separating the cleaning area from suction zone and thus avoids losses in cleaning efficiency.

The delivery ducting is skilfully designed so as to suck the material off the spikes of roller B. For this, the bending of the delivery duct is such that it helps the suction to draw an air almost tangentially to the roller B. It is evident that as the speed of the suction helps in increasing the production; it also affects the cleaning carried out within the machine.

5.3.5.7 Multi-Function Separators[7]

As the name suggests, various different functions such as heavy part separation, metal detection and its separation, fire sensing and its extinguishing are all carried-out in one extended machine. All these become very important, especially when finally the material has to be fed to the card where the striking elements are much finer and delicate.

The normal position of the machine necessitates that it be placed as early as possible in the blow room. Therefore, in modern blow room, the machine is arranged as the second (or maximum third) machine placed immediately after Automatic Bale Opener – Blendomat. The material in the form of small fibre-tufts is received by this machine (Fig. 5.40) from automatic bale opener. It is deposited on a cage. The suction fan draws the air through the cage. It thus helps in bringing the material on to the cage.

The subsequent journey of the material through the machine is owing to another suction provided after the delivery end from where it is suitably guided to the next machine. The separation of heavy particles takes place at A where they are separated due to gravitational effect. The material is then directed through pipe B. At C, the material is made to pass through a perforated tube, the perforations helping the fine particle separation under the influence of air suction at D.

During the onward journey, the material is scanned by a metal sensor and when any metal particle is detected, it is effectively removed by an electronic

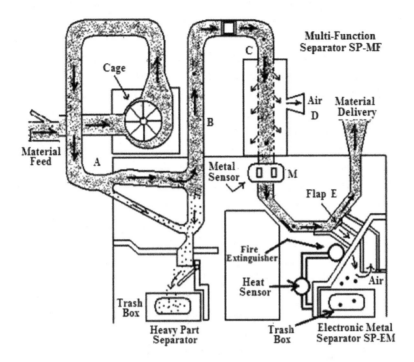

FIGURE 5.40 Multi-Function Separator SP-MF[7]: Various different functions are incorporated in this machine. For example, there is aero-dynamic separation of lint & trash; there is metal detection and its electronic separation and heat sensing &fire extinguishing.

metal separating device. The journey-time between M and E is electronically computed and the flap E opens out after a calculated time lag. Here again, the path of the material is such that the ejection of metal particle is brought about by the aero-dynamic principle. The heat sensing device is connected to the trash box and in the case of any emergency, the signals are passed-on to fire extinguishing device. Finally, the material is let out through delivery chute to the next machine.

The bar chart (Fig. 5.41) brings out the energy consumption aspects. Both the power required to operate the machine and that required for the suctions at various places for the filters, is comparatively much less. It is thus, highly recommended that the Multi- Function Separator unit should form the integral part of any modern blow room line. A typical Trutzschler blow room line (Fig. 5.43) invariably includes this machine. The sequence is shown in Fig. 5.42.

The production of the machine can be as high as 1500 kg/h, and it perfectly matches with the out-put of Blendomat. With the control over the fan speed (speed through the machine), the separation of heavy particles, especially seed-coat fragments, is achieved at a high efficiency level. At the same time, there is considerable saving of good fibres. The main airflow is used to carry the material to the next machine, whereas only a small portion of air (600 m^3/h) is used for filters where the impurities are extracted.

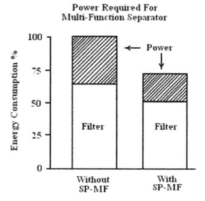

FIGURE 5.41   Energy Consumption[7]: It reveals that there is considerable power saving when SP-MF is used.

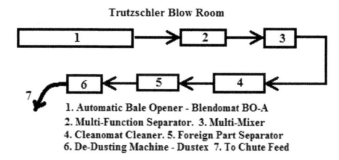

FIGURE 5.42   Modern Trutzschler Blow Room Line[7]: The speciality of any modern blow room line is that it uses minimum number of machines carrying out the job of opening & cleaning; and yet gives excellent cleaning performance.

5.3.5.8   Securomat[7]

Even though Securomat resembles in some respects to foreign part separator, it has a distinguished feature in identifying, locating and removing the trash particles.

The special cameras (Fig. 5.43) are used to take the pictures of the working surface of the opening roller continuously. The mechanism is linked with an assembly consisting of air jets, which are installed to blast the fine air jets on the material web about to leave the fine pins of opening roller. There are several jets and each of them is activated by the signals received from the camera images. After the impurities are identified and located, the signals from these cameras are conveyed to only specific jets. Only those jets which are concerned with the area of impurities across the fibre web are thus activated. The jets from the corresponding nozzles then blast away the impurities. For enabling the cameras to take continuous photographs of the opening roller surface, there are four fluorescent tubes which constantly illuminate the opening roller surface. It is, thus, possible to identify the impurities against the roller surface. This is because; the reflectance of such

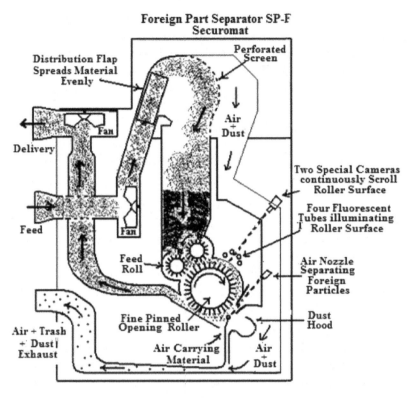

FIGURE 5.43 Securomat Principle[7]: A novel principle is used in detecting the foreign particles by two special cameras and then removing them by fine air-jets. The dust, as usual, is separated by suction through perforation.

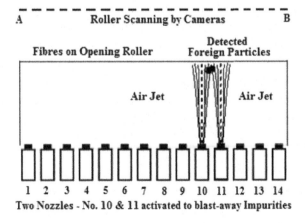

FIGURE 5.44 Air Nozzles[7]: The appropriate and specially designed nozzles sent the blast of air at proper time to forcibly remove the foreign body detected by camera.

impurities is quite different from the normal fibrous web. The air nozzles are so situated that their blasting air jets drive the impurities tangentially and away from the roller surface into a zone where the air currents, not very strong, are induced to suitably carry these trash particles into a dust hood. There is also a provision for the gathered impurities in this hood to be suitably taken away.

As seen from the figure (Fig. 5.44), at a particular instant, only jet number 10 and 11 are activated to drive away the impurity in the concerned zone.

5.4 MODERN BLOW ROOM LINE & RELATED RESEARCH

5.4.1 TYPICAL MODERN BLOW ROOM LINES[7]

With the use of Modern concept and machinery, the modern blow room appears to be far more compact as compared to conventional blow room. The earlier concept of using 7–9 openers/beater in a line and giving heavy beating action so as to extract impurities has long been forgotten.

In old blow room, some of the beaters like Crighten Opener were very harsh in their action and a by-passing arrangement had to be provided so as to avoid harsh action when processing finer cotton varieties. The modern line hardly contains 2–3 types of openers, and those too, mainly concentrate on giving intensive opening action and avoiding harsh beating action.

The new machines like multi-function separators, (Fig. 5.45) dust extracting units, machine like CVT cleaners completely replace the old conventional machines. Especially action of CVT type of machines (Fig. 5.46) is unparalleled; in that it is not

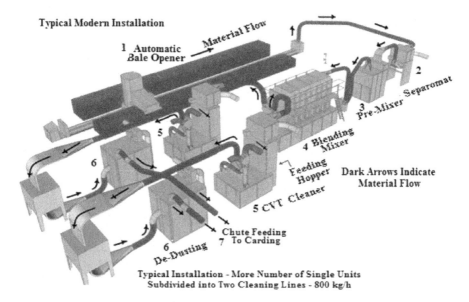

FIGURE 5.45 Modern Blow Room Sequence[7]: Automatic bale Opener is the keyword for all modern blow room lines. Very high cleaning efficiency of such lines is mostly due to very effective reduction in the tuft size.

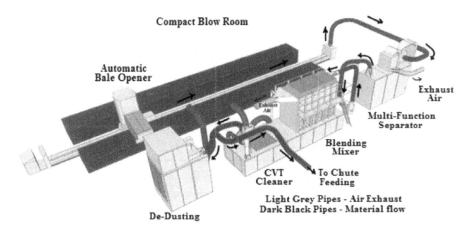

FIGURE 5.46 Compact Modern Blow Room Sequence[7]: The efforts are always made to introduce more effective opening & cleaning machines in modern blow room lines. This makes it possible to reduce the number of machines in modern blow room sequence.

harsh and yet is very effective in thoroughly opening the cotton tufts to the finest extent.

This, on the one hand, improves the cleaning efficiency remarkably, whereas the fine opening of the tufts makes the work of the subsequent carding operation much easier. The multi-function separator, along with extraction of fine dust, also provides safety for any consequential fire within the line.

The magnets provided therein do not allow any metal particle to escape further and thus avoid both sparking and damage to the metal parts of the machines. Dust extraction units try to further improve cleanliness of the material.

5.4.2   MODERN CONCEPTS IN OPENING[8]

In technological sense, 'opening' means reduction in specific density of cotton flocks. In short, there is an increase in the volume for a given weight of flock. There are many methods such as 'Fractionator', 'Porometer' or even 'Weight & suspension of the flock by open-can method' available for finding the openness of cotton.

The basic principle involved in such methods is to find the compressibility; the more the flock is compressed, the higher is its openness. There are always two stages in opening action. First, the tufts are opened to reduce the specific density. Second, they are broken into more than one smaller tuft. Whereas breaking into smaller tufts would suffice the purpose of cleaning, the opening of the tufts followed by the reduction in specific density is essential for blending and even for obtaining the regularity at the final blow-room stage and thereafter at subsequent stages. Especially, in chute feeding to the card, inadequate opening of the tufts is most likely to lead to shortening of staple. Thus, to know a correct evaluation of 'Degree of Opening', both the size of the flock and the measure of its density (g/cm^3) would be needed. However, in practice, a mass in mg/flock usually is more popular as it is easy and quick to find.

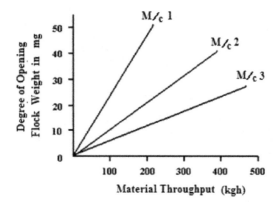

FIGURE 5.47   Degree of Opening[8]: The opening of the tuft is seriously affected by the rate of throughput. The performance of the machine is better if rate of increase in the flock weight is slower with increase in this rate of throughput.

It can be realized that the rate of the production of any machine is going to affect its performance in terms of its opening action. Thus, when the production rate is increased, (Fig. 5.47), the action of opening of the tufts in any machine would be reduced.

Such graphs help in choosing the machines on the basis of their performance. Amongst the three machines, the performance of machine No. 1 seems to be ser- iously affected in terms of its opening action (flock weight).

With increase in production, there is a rapid increase in the flock weight in this machine which leads to poor opening. As against this, in machine No. 3, this opening action is much less affected. It continues to give far better opening action even when production rates are increased to almost five times.

Similarly, it can also be seen (Fig. 5.48) that the performance of machinery sequence used in a typical blow room line in terms of flock weight reveals the

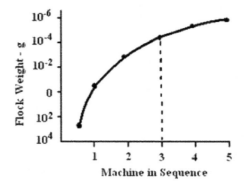

FIGURE 5.48   Degree of Opening[8]: In a blow room machinery sequence, each machine has to steadily continue opening the tufts. If any machine does not follow this requirement, it becomes redundant.

efficacy of their opening action. Whereas on x-axis, there are machines in sequence, on y-axis, it is the flock weight in decreasing order. It is revealed that up to machine No. 3, the opening of the flock (flock weight) is steadily continued. Machine No. 4 and No. 5, however, do not seem to contribute much further to the opening action. This can be seen from the flattening of the graph thereafter.

It is, thus, very important in choosing a sequence to see that each subsequent machine continues to contribute to the increase in the openness of the tufts progressively. With reference to above graph, machine No. 5 may be considered to be really superfluous. Any additional machine of this kind not only involves an extra cost (& space), but also it may unnecessarily stress the raw material without helping the opening action. A possible use of machine No. 5 can only be justified, if it helps in improving subsequent carding action, assuming that it is the last machines in blow room sequence before chute feeding to the card.

The intensity of the opening that the cotton receives at every blow-room machine stage, depends upon how the beaters are clothed. In this respect, the needles or pins or even saw tooth wire would have maximum opening action. However, in blow room, their cleaning power would have limitations. In addition, the opening power of the machine would also depend upon the speed of opener or cleaner and its production rate. Thus, with higher speeds of openers/cleaners and lower production rates, the opening action is expected to be improved.

5.4.3 MODERN CONCEPTS IN CLEANING[8]

The cleaning action in any machine is carried out by giving a quick and fast movement to the cotton flocks and striking them against the grid bars. There are various different actions in the machines in blow room sequence which take place at different points, e.g. the impurities made to fall out through grid bars, the development of the centrifugal force to beat the cotton and eject the trash and rubbing of the cotton tufts against metal parts to separate the impurities.

When the cotton tufts are repeatedly struck against the grid bars, cleaning takes place and the impurities fall-out. The beating with severe blows causes comparatively harsher action on cotton tufts. The force and the suddenness of the blow with which the impurities are struck result into their removal from the cotton stock. This is also possible because of the high acceleration of impurities and their comparatively smaller size. When the tufts, after having been opened are made to brush against the grid bars, knives or grate, it results into separation of the impurities entrapped in the flock. In addition to this, in almost all modern blow room machines, the suction is beneficially used in extracting the fine dust. This is carried out when the flock is transported over a perforated or wire mesh. The opened condition of the cotton stock itself is responsible to result into separation of the impurities and the rest is done by air suction.

In every blow room sequence, it is comparatively easy to remove the larger, coarser and heavy dirt particles in the early stages. In this case, if possible, cleaning should immediately follow the opening within the same machine. The higher speeds of cleaning rollers always lead to better cleaning action; however, there is always more stress on the fibres. Also, the cleaning action does not significantly improve

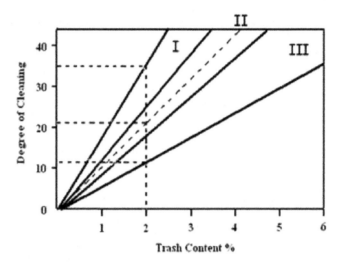

I. Low Resistance Zone II. Medium Resistance Zone III. High Resistance Zone

FIGURE 5.49 Cleaning Resistance[8]: Depending upon the holding power of the cotton for trash, the cleaning takes place at every cleaning machine. Some varieties of cotton resist this cleaning action more than the others.

above a certain speed of the cleaning element. In such case, only the stress on the fibres continues to increase. It is also accompanied with more fibre loss.

5.4.3.1   Resistance to Cleaning[8]

When dirty and trashy cotton is mixed with comparatively clean cotton, the cleaning becomes difficult. The cleaning efficiency of the machine depends on the trash content; the size of the particles and how tenaciously they cling or adhere to the fibres. There are some types of cotton to which the impurities adhere more strongly. Thus, the cleaning efficiency of a machine may be different for two different cotton varieties even when their trash content is the same.

A new concept called 'Cleaning Resistance' is introduced to represent the adherence of impurities to the fibres.

It can be seen (Fig. 5.49) that the cotton belonging to 'low resistance' zone (I) has much better cleaning level, e.g. for 2% trash content, the degree of cleaning achieved by the cotton in this zone is around 35%; whereas that achieved by the cotton in 'high resistance' zone (III) is only 12–13%.

The cleaning ability of a machine, therefore, can be described as functions of:

(1) Machine Performance Factor (M) (2) Cotton Properties – Factor (C) and (3) Trash Content (T) in the following way:

Cleaning Efficiency [C.E.] = a. M. C. T. (where 'a' is constant)

Usually, the values for both 'M' and 'C' vary through the same range (0.5–1.5).

The typical curve in Fig. 5.50 is for three machine set-up. In the region of values 0.5 and 1.5 for C-factor, the cleaning efficiency levels of 40–65% are obtained. For values less than 0.5 or more than 2.5, the influence of material is more.

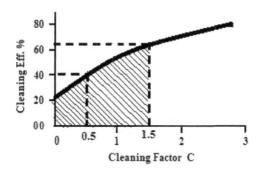

FIGURE 5.50 Influence of 'C' on cleaning efficiency[8]: There is always the best cleaning opportunity for the cleaning machines when the 'C' factor lies between 0.5 & 1.5.

Thus, if the raw material is to be cleaned to its optimum level, the knowledge of the material properties will have to be better known. In the same manner, the following two graphs will give a fair idea regarding the ability of the machines to carry out the cleaning.

Fig. 5.51 is for old conventional blow room sequence, whereas Fig. 5.52 is for a modern blow room plant. In both these sequences, the same quality of cotton was processed. The machines X_1 & Y_1 are the starting machines in the respective sequence. In the conventional sequence X_2 to X_5, the machines attain around 40% 'cleaning efficiency'; whereas in modern sequence Y_2 to Y_4 (only 3 machines), attain around 55% 'cleaning efficiency. It may be noted that the graph (Fig. 5.51) after the machine X_3 really does not show significant rise in cleaning efficiency with the conventional sequence. Whereas, there is a steady increase in cleaning efficiency up to Y_3, and beyond it, (from machine Y_3 to Y_4), a pronounced rise is seen for modern blow room sequence (Fig. 5.52).

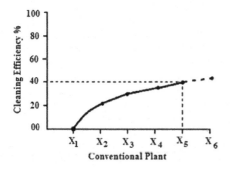

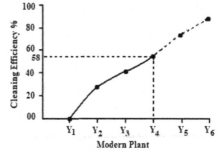

FIGURE 5.51 Cleaning efficiency[8]: When a progressive opening is maintained, a comparatively higher cleaning efficiency is achieved with only a few machines in the blow room sequence. This is an important feature of modern blow room.

FIGURE 5.52 Cleaning Efficiency[8]: When a progressive opening is maintained, a comparatively higher cleaning efficiency is achieved with only a few machines in the blow room sequence. This is an important feature of modern blow room.

As stated earlier, the conventional sequence does not focus its attention on continued tuft reduction both in size and weight. The later machines in the sequence, therefore, are not able to add much to the cleaning level achieved earlier. However, in modern sequence, the stress is mainly on continual reduction in tuft weight and its size. This automatically improves cleaning carried out at every stage in the sequence. The points X_6 and Y_5 or Y_6 in the respective graphs are the possible additions of machines in the corresponding sequences. Here again, it can be seen that, whereas an additional machine in conventional set-up would only marginally increase cleaning efficiency, an additional machine (Y_5) like intensive opener would further reduce the tuft size and weight and thus would enable a marked rise in cleaning efficiency level.

It may be mentioned here again, that the cleaning achieved in the blow room line does reduce a burden on card, which has to carry the same function further. However, the main function of carding process being individualization of the fibres, the level of openness of cotton tufts achieved in blow room would not only substantially improve the individualization but also automatically improve cleaning carried-out at card.

5.4.3.2   Other Considerations[8]

Though the cotton contains comparatively less amount of dust before ginning, new dust is created during the processing owing to shattering and smashing action of the blades, spikes, pins and saw-teeth of cleaners. The dust particles thus become very light and almost float with the fibres. Further, when associated with fibres, they adhere to them more strongly. Therefore, they need to be rubbed-off. Thus, the dust can be eliminated in the process by high metal-to-fibre or fibre-to-fibre friction.

It is known that the degree of cleaning cannot be better than the degree of opening. Hence, each subsequent machine in any sequence should strive hard at improving the opening action. Further, it is also known that the impurities can only be removed from the surface of the tufts. Therefore, new surfaces must be created at every machine in the sequence. The cleaning too, should be simultaneously carried out with every opening action. It may be noted that a high degree of opening achieved through harsher action is likely to reduce the staple of cotton.

5.4.3.3   Rieter's Vario-Set Concept[7]

The cost of the raw material constitutes almost 50% of the total cost in a spinning mill. Therefore, restricting good fibre loss through Blow Room and Carding becomes very important. One of the basic functions in both these departments is cleaning of impurities from cotton. It can, therefore, be easily realized that any efforts made for extracting the trash or the impurities from the bale cotton would necessarily involve some loss of lint (good fibres). It means that, whenever the cleaning operation is intensified, the proportion of the lint in the droppings would increase. This leads to the concept of optimum cleaning.

It is evident from Fig. 5.53 that when the degree of cleaning is increased (intensified), there is substantial good fibre loss as seen in the blow room droppings.

This means that any effort to extract more trash from the bale cotton in the blow room not only increases the total droppings in the blow room; but also these droppings are richer in their value (more lint%). The graph also reveals that harsher treatment increases the neps sizably, especially when processing cottons with higher immature fibre percentage. It is needless to state that the harsher treatment is very likely to cause fibre damage.

With a normal blow room working, the action of any single machine or group of machines, when intensified, leads to more extraction of trash (more droppings) and therefore gives a comparatively cleaner delivered material.

But this cleaning is at what cost (good fibre loss) will have to be ascertained especially if the overall material belongs to better class. The technicians have to frequently watch the nature of droppings and make timely corrections in the settings.

Rieter has provided a Vario-Set concept where the speeds and setting can be chosen so that it not only gives the desired amount of droppings but also a pre-determined composition of the droppings in terms of trash and lint (Fig. 5.54 a, b). It is possible to define the quality of the blow room operation in terms of sufficient cleaning, much-less good fibre loss, minimal fibre damage and more control on nep generation.

The term 'sufficient cleaning' decides what should be passed on to the card. However, the term 'cleaning' itself is relative. Cleaning fine dust would be more beneficial to rotor spinning, whereas Air jet spinning would appreciate trash-free feed material. With fine fabrics and knit-garments, nep-free yarn may enhance the fabric quality, as against this, fibre damage may be of great importance in ring spinning.

As a modern plant has a direct chute feeding arrangement to the card, the blow room line for any reason, cannot be stopped for carrying out any changes. In earlier

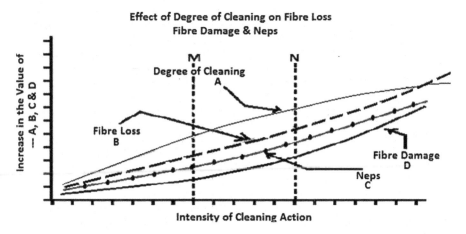

FIGURE 5.53 Intensifying Degree of Cleaning[7,8]: When it is tried to intensify cleaning action, only the degree of cleaning is improved. However, the fibre loss, the fibre damage & the neps – also increase, thus deteriorating the quality.

sequence, especially when making the change for increasing or reducing the intensity of the blow room machines, the whole line was required to be stopped.

The settings of beater blades with mote knives were then carried out. The speeds of the beaters were altered by changing the driving pulleys. The blow room could be started only then. During the stoppage, the lap making process (in conventional blow room line) was also stopped. But with sufficient stock of reserve laps, there was not much loss of card production. With the chute feed sequence, however, stopping the blow room line is out of question as it directly halts the production of all the cards linked in a chute feed system.

The ingenuity of Vario-set concept is that while making the changes in the settings to optimize the lint and trash proportion in the droppings, it is not necessary to stop the machines. The blow room continues to work and so do the inter-linked cards. There are four major areas (working points) in Vario-Set field – A, B, C and D (Fig. 5.54a). These areas represent different proportions of lint and trash in the blow room droppings (Fig. 5.54b). They offer different options for choosing the intensity of the action of cleaners used in Rieter's line. As can be seen from In Fig. 5.54 (a), A-zone is more trashy, whereas D-zone contains very negligible trash and has more proportion of lint. It is not necessary to go physically for changing the actual setting of cleaning elements (grid bar or cleaner-speed). Whatever composition of blow room (or individual machine) dropping in terms of lint and trash is desired, the corresponding buttons on the screen (Fig. 5.54a) can be just pushed. The change in the speeds and setting the grid bar positioning is automatically carried out. As mentioned, all this can be done when the machines are running. This saves production time.

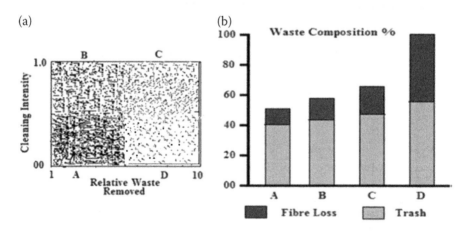

FIGURE 5.54 Vario-Set Field[7,8]: The effect of increasing grid bar angle (opening) and the cleaner speed affects the percentage of lint & trash in the droppings (waste removed). Both or any one of them can be changed to suitably arrive at the acceptable proportion of lint in the trash: (a) Working points inVario-Set Cleaning Field; (b) Proportion of Lint & Trash with – Vario-Set Points.

Grid Bar Angle

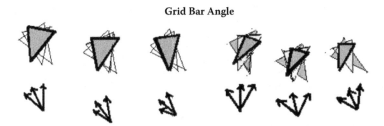

FIGURE 5.55 Automatic changing of Grid Bar Angle[7,8]: This is linked with vario-set field. Once a certain area of desired proportion of lint and trash is decided, automatically the grid bar angle (grid bar opening) gets adjusted.

It can be realized that, with every angle of grid bars, there is a certain intensity of cleaning action linked. When the angle is changed, so is the distance of bars from striking element. So also with the change in the angle, the direction of the sharp edge of the grid bars facing the beater blades is decided. Therefore, the beater speed as well as the grid bar angle both decide the intensity of action. In fact, a combination of a particular beater speed together with a certain grid bar angle can give many possibilities (Fig. 5.55) of levels of cleaning intensities. The buttons on the screen of Vario-Set field automatically alter the positions of the grid bars underneath the openers. The composition of the droppings as lint and trash is thus automatically decided and set. To a certain extent, the decision of selecting the area on Vario-Set field depends on the nature and quantity of trash in the material fed (bale cotton). The technologist has to bear in mind that if any extra efforts are made to clean the cotton that is comparatively cleaner (trash content < 3%), there is bound to be large proportion of lint in the droppings. Depending upon the quality and the economics of the cotton material processed, how much intensified should be the cleaning treatment and at what cost, can be arrived at.

The bar chart (Fig. 5.54b) shows the composition of the droppings, i.e. the relative quantity of lint (good fibres) and the trash. The chart reveals that when the settings are steadily moved from A to D, the droppings become more whitish. This means that the proportion of the lint in the droppings is alarmingly high.

Depending upon the degree of contamination, the Rieters provide the sequence of machines. It must be remembered that the production rate required also governs the sequence. Taking these together, Rieters have suggested the following sequences:

A-11 – Flock Feeder, B-12 – Uni-Cleaner B-17 – Uni-Cleaner, A-79 – Uni-Store, B-72 – Uni-Mix (for 800 kg/hr), B-76 – Uni-Mix (for 1200 kg/hr), A-81 – For precise metering and small weight ratios of blend components Rieters have 4–5 machines in their modern blow room sequence (Fig. 5.56) to cope-up with varying trash content in the material from slight contamination to high degree contamination.

With the by-passing arrangements provided for important machines, the sequence for low, medium or high trash content in the material fed gets automatically

TABLE 5.2
Machine Sequence for Low Trash Content

Production Rate	Machine Sequence			
800 kg/hr	A-11	B-12	B-72 R	
1000 kg/hr	A-11	B-12	B-76 R	
1200 kg/hr	A-11	B-12	B-76	A-79 R

TABLE 5.3
Machine Sequence for Average Trash Content

Production Rate	Machine Sequence – Average Degree Contamination			
800 kg/hr	A-11	B-12	B-72 R	A-79 R	
1000 kg/hr	A-11	B-12	B-76 R	A-79 R	
1200 kg/hr	A-11	B-12	B-76	A-79 R	A-79 R

TABLE 5.4
Machine Sequence for High Trash Content

Production Rate	Machine Sequence – High Degree Contamination				
800 kg/hr	A-11	B-12	B-72 R	B-17	A-79 R
1000 kg/hr	A-11	B-12	B-76 R	B-17	A-79 R
1200 kg/hr	A-11	B-12	B-76	B-17	A-79 R

selected without wasting time. In every machine (or sequence), the emphasis is given on a gentle cleaning and opening. Thus, the fibres experience only as much stress as is absolutely essential for blow room treatment. This also avoids the reduction in the fibre staple (less damage) and thus helps in reduction of neps generated in the blow room.

5.5   CLEANING EFFICIENCY & FIBRE LOSS[3]

A greater opening of cotton stock in the blow room, however, does not always mean greater cleaning, e.g. in SRRL, a very effective opening action is possible. But the cleaning efficiency of the machine is not so high. This possibly is because there are no adequate provisions within the machine to separate trash effectively from opened-out cotton lint. Usually, a greater cleaning level through higher speeds and close settings

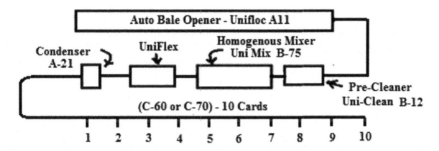

FIGURE 5.56 Modern Blow Room Line (Rieter)[7]: The trend has been set to directly feed the blow room material to the card – called chute feed to Card. This linking solves many problems – both economical & technical. **Modern Single Process Blow Room with Chute Feeding to Card.**

of beater elements with grid bars, always leads to high lint loss. This is also the case with many other cleaning units. A spinner should know the proportion of lint in the droppings under each machine in the sequence as well as tolerable lint loss in the blow room droppings as a whole. An equation can be derived for the same as follows.

Let,

x = Trash% in the bale cotton; y = Trash% in the lap & z = Total blow room droppings.

1. With 100 kg of bale cotton processed – $(100 - z)$ kg will be forming laps.
2. This lap material would contain – $(100 - z)$ $y/100$ kg of trash in it.
3. Hence the weight of trash in the blow room droppings would be –
   = Trash in Bale Cotton – Trash in Lap = Actual Trash in the Droppings.
   = $[x - (100 - z) y/100]$ Trash by weight in the droppings.

4. Expressing this as a percentage of the total droppings (z), we get,

$$\text{Trash Content in BR droppings} = \frac{[x - (100 - z). \, y/100]}{z} \times 100 \qquad (A)$$

The proportion of the lint in the blow room (BR) therefore will be –

5. Lint Loss % = $[100 - (A)]$

The values of 'x', 'y' and 'z' can be easily obtained by using Shirley Analyser. So also, the data for total blow room droppings can be obtained by collecting the droppings over a shift. This, however, is not possible in modern blow room line. It is known that the total blow room droppings are related to the trash content in the material fed.

Thus generally, if the trash content in the bale cotton is 'x' then the total droppings should not exceed $(x + 0.5)$ for low trash levels and $(x + 1.0)$ for higher trash levels in the material fed. It can be seen from Table 5.6 that as the amount of trash in the cotton increases, the cleaning efficiency of the blow room also increases without intensifying blow room action. But this increase is not proportionately. This can be seen from Fig. 5.57a.

With higher trash in the cotton, the heavier types of impurities are much easily and quickly removed. The dropping percentage, therefore, increases (Fig. 5.57b). However, with better grade cottons, the trash is comparatively lighter in nature and it becomes more difficult to remove it.

When the blow room treatment is intensified, especially with cottons containing lower trash, it is always accompanied with higher droppings and fibre loss (Fig. 5.58). Therefore, as the removal of trash is always accompanied, to certain extent, by the lint loss, the object of opening and cleaning should be to minimize this loss.

This is because with the greater proportion of lint in blow room dropping, the good fibre loss will be sizable. Normally, when the trash content in the bale cotton is higher, the corresponding lint loss is considerably lower.

Usually, there is comparatively higher lint loss (Fig. 5.58) with cleaner cotton. The lint is generally lost through the spaces between the grid bars and, hence, their

TABLE 5.5
Cotton Grades Based on Trash Content

Trash Content%	Cotton Rating
1–2	High Grade
3–4	Medium Grade
5–7	Low Grade
Above 8%	Very Low Grade

TABLE 5.6
Dependence of Blow Room Cleaning Efficiency on Trash in the Mixing

Trash Content%	Cleaning Efficiency
1–2	52–58
3–4	58–68
4–5	68–75
5–6	75–78
6–7	7–80
Above 7	80–84

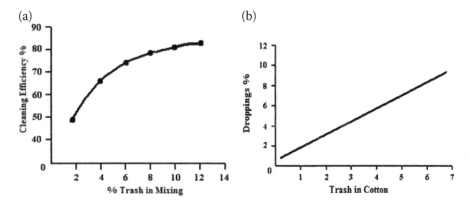

FIGURE 5.57   Cleaning Efficiency & Blow Room Droppings[3,8]: It is evident that more trash in the mixing gives more opportunity to the cleaning elements to strike and extract them as waste. This improves cleaning efficiency. As a result, the total droppings increase and therefore their percentage increases: (a) Trash in the Mixing v/s Cleaning Efficiency[3]; (b) Trash in Cotton v/s Blow Room Droppings[3].

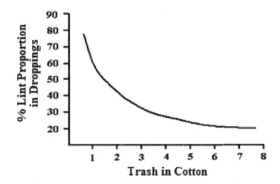

FIGURE 5.58   Lint Loss in Blow Room[3,8]: With low trash content, the lint easily follows the trash through grid bars, With higher trash content in cotton, it is the trash which gets priority to be ejected. Hence, the lint percentage reduces.

correct positioning and setting are very important. The use of more number of grid bars or closing the gap between them (changing angle) can reduce the lint loss.

5.5.1   RECOMBINATION OF LINT & TRASH[6]

When the material travels through a given sequence of machinery in blow room, it is assisted in several ways – feed lattices, pneumatic means or gravity. With pneumatic transportation, many a time, the pipes conveying the cotton material extend over a longer distance. In such cases, it is often experienced that the loosened out and yet unremoved impurities in the earlier machine have a tendency to reunite with cotton lint. This takes place either during transportation or when subsequently the material is collected in the hopper.

This is mainly because the earlier machine fails to take full advantage of separated-out trash and does not extract it fully. Thus, the part of the loosened-out trash and cotton lint travel together in the long, extended pipes which are occasionally bent. The result is obvious. The lint and the trash find ample opportunity to get re-combined.

This was found out by comparing the cotton tufts for their trash content,[6] immediately after they were delivered by the machine (A) and when they subsequently reached the feed side of the next machine (B). Normally it is expected that the trash content at (B) would be comparable with that at (A). However, when the cotton tufts were analysed for their trash content, it led to anomalous results indicating that trash content (B) was higher than that at (A). This was because the material collected at (A) was in free flow and, as such, it was basically the freed lint from the trash. During the passage from (A) to (B), both the freed lint and the accompanying trash had plenty of time and chances to recombine with each other. It may be noted that this recombination does not again allow the trash to be entrapped within the cotton tufts as both travel in a free flow. The trash in all possibilities is expected to be merely over the surface of such tufts.

It is observed that lighter and smaller tufts show more tendencies to recombine with trash. This is basically thought to be owing to their comparatively larger surface area. It is possible to avoid this recombination by improving the cleaning action at every machine. So also, it is possible to avoid using long conveying pipes. Whenever such longer distances cannot be avoided, some means of aerodynamic cleaning can also be introduced in between, so that most of the trash gets separated again and extracted. This is very likely to minimize the chances of recombination of lint and trash.

In modern blow room machine sequence, the total number of machines in a sequence is very small. Each machine, when concentrating on opening the cotton tufts, also simultaneously focuses its attention on cleaning. The degree of opening the tufts in each succeeding machine is continuous and progressive. Hence, it is possible to carry very effective cleaning. It may be mentioned here that the machine action on cotton in the modern blow room is comparatively gentle as against harsher action caused by some typical machines in conventional blow room. Therefore, even with fewer machines used in modern blow room sequence, it is quite possible to achieve betters cleaning efficiency levels. In addition, better opening in the blow room followed by a chute feeding improves the performance of a modern card, which is fully equipped with far better contrivances around it to improve the cleaning efficiency further. As a result, the overall cleaning efficiency with the modern set-up (blow room and card taken together) is really excellent (>97%). This leads to much-better cleaned sliver and hence a superior quality of yarn. Such high levels of performances are not possible in conventional set-up.

LITERATURE REFERRED

1. Introduction to Study of Spinning – W. E. Morton, Longmans Green & Co., 1952.
2. Manual of Cotton Spinning – "Opening & Cleaning" – Vol II, Part II – W. A. Hunter & C. Shringley, The Textile Institute Manchester, Butterworths, 1963.
3. Process Control in Spinning, ATIRA publications, 1974.
4. Book of papers, Training Programme for Research Persons, Nodal Center for up-gradation of Textile Education, Seminar of Blow Room & Carding, I.I.T. Delhi & V.J.T.I., Mumbai.
5. Technology of short staple spinning – W. Klein.
6. A Blow Room practice – Pierce, Kelly & Kolman, JTI 1955.
7. Brochures, pamphlets & booklets of Rieter, Trutzschler & Trumac.
8. Manual of Textile Technology – W. Klein, The Textile Institute, Manchester, U.K.

6 Defects in Blow Room Product & Machinery

6.0 DEFECTS[1-3]

Basically, many of the following defects are related with lap-forming process associated with conventional blow room. In modern blow room line, the material from last machine in the blow room sequence is directly fed through chute to the Card. In India, especially, the rate of change-over from conventional blow room line to a fully modern blow room line with chute feeding has been very slow. There are some mills, still working with conventional set-up. Even then the concepts are applicable in a different sense, e.g. uneven feeding to a card is applicable both in conventional as well as modern set-up. The rich droppings or lower blow room cleaning efficiency is also applicable in modern set-up.

6.1 UNEVEN LAPS

The lap formed in blow room is expected to be uniform along its length and width. Normally uniform feeding from feed regulating motion is expected to ensure that feeding to the beaters during the equal time interval remains fairly constant. Similarly, the degree of opening has a profound influence in maintaining a uniform distribution of cotton across the cages. It also controls and maintains the uniform density of cotton issued from the reserve boxes. The variation in the lap may be of short term or long term nature. The probable causes are as follows.

- Uneven feeding of cotton.
- Faulty working of feed regulating mechanism.
- Shifting of belt over the cone drums not smoothly taking place – either the belt has become too stiff (not pliable) or it has not been pieced-up properly. A belt strap wider in width is not sensitive enough to bring about the changes sensed and conveyed by the regulating mechanism.
- Uneven loading on the feed roller and improper alignment of the feed roller with the beater – allow uneven snatching of fibres. The slippage of pedal roller causes uneven delivery of material.
- Feed lattice slipping over its driving shell.
- Too low fan speed – Easy and smooth withdrawal of the material through the beater zone not allowed. The air turbulence due to dirty flue pipes and cages affects uniform distribution of cotton on cages.

6.2   CONICAL LAPS OR BARREL SHAPE LAPS

If the dust box is too small, the air entering the cage perforations and subsequently being led into dust box experiences backpressure. This leads to uneven distribution of material over cage surface. Similarly, if the cage perforations are partially (or fully) choked-up on any particular portion of the cage, it will draw less air through those sections and hence will result into thinner deposition of the fleece.

- Some of the pedals in feed regulating motion jammed, thus not effectively conveying the variation in the thickness of the cotton sheet.
- Uncontrolled airflow through air in-let valve placed on either side of the trash box – allowing uneven drawing of air through the cages.
- Blunt beater blades resulting in uneven striking action and thus withdrawing uneven cotton material from the feed system.

6.3   DIRTY LAPS

The laps contain substantial amount of visible impurities and thus appear dirty. This indicates poor cleaning in blow room.

- Grid bars are too close to each other. They are required to be partially open for the liberated trash to fall down through them.
- Excessive fan speed preventing cotton from being adequately cleaned in the beater zone.
- Beater speed is comparatively slower, causing ineffective cleaning action.
- Setting between the beater blades and the pedal roller too wide, thus causing lumps of cotton being snatched.
- The mild air currents returning back into beater chamber through the grid bar causing the trash to return back into beater zone – recombination of lint and trash within the beater.

6.4   STRINGY LAPS

Overbeating action of some of the beaters in blow room curls the fibres. A tuft of fibres that is repeatedly beaten up by the beater blades tends to form 'strings'. With heavier beating, the string size also grows. Once formed, the strings are very difficult to open out in subsequent processes and may cause difficulties in drafting process. The possible reasons are: (1) Too high beater speed (2) Too low fan speed and (3) Excessive moisture in the cotton processed.

6.5   RICH DROPPINGS

It means that there is considerable proportion of lint or fibrous matter in the droppings under the beater. The possible causes are –

- Excessive amount of soft waste in mixing.
- Grid bars set either too deeply and/or are more open.
- Air suction of the fan too weak, thus allowing cotton to linger in the beater chamber.
- Excessive beater speed leading to more forceful action on the material being processed.

6.6   DEFECTIVE LAP EDGES

They usually occur due to accumulation of waste in the holes at the sides of the cages. This weakens the air suction on the sides of the cages, leading to irregular deposition. It leads to torn or ragged edges. Also, it is quite likely that some roughening on the inside surface of the passage that leads to cages cause intermittent deposition. An inspection can help in locating the roughenings. A periodic cleaning of the cages helps in preventing these blockages of the perforations on the cage.

6.7   LAP LICKING[1]

This trouble is experienced only when the lap produced in blow room is put on the lap roller of the card. During unwinding of lap at card, the adjacent layers stick to each other. However, due care needs to be taken during lap building itself, in the blow room.

- The mixture of comparatively long and short fibres is very difficult to be consolidated during the lap formation and hence there is tendency for lap licking.
- Incorrect proportion of soft waste makes the laps more fluffy.
- Atmospheric conditions are too humid or damp-cotton is added in mixing. This increases the tendency of fibres to adhere to each other, thus leading to cross-layer fibre adhesion which is licking.
- Unequal amount of cotton, when made to deposit on the top and bottom cage, reduces this tendency. During calendering, the use of felting fingers or scroll roller (the latter making indentation marks) becomes more effective. Forming lighter laps, uneven diameter of top and bottom cage, use of roving to separate the layers, more effective use of dampers to divert unequal material to deposit on the cages – also reduce the lap licking tendency. Some of these are required to be followed, especially while processing man-made fibres.
- If the cage runs too fast, it leads to improper blending of fibres during their deposition on the cage. This may also cause licking.
- Inadequate calendaring pressure affects the compactness of the lap. Whereas too high rack pressure on the lap spindle increases the pressure on the neighbouring layers to come closer and stick to each other.

LITERATURE REFERRED

1. Cotton Spinning – William Taggart.
2. Manual of Cotton Spinning – "Opening & Cleaning" – Vol II, Part II – W.A. Hunter & C. Shrigley, The Textile Institute Manchester, Butterworths, 1963.
3. Elements of Spinning – Dr. A. R. Khare.

7 Blow Room Calculations

7.1 BASIC INFORMATION[1]

Knowledge of how the size of a particular yarn is defined is the first essential thing in yarn or cloth calculations. In the case of coarse metal wire, it is customary to specify the diameter of wire either in 'thou' (1/1000 inches) or in mm. In the case of a still finer wire smaller units may be used. This is called the gauge of the wire. However, it is very difficult to get worthwhile information about the size of the yarn from its diametric measurements. This is because the diameter of the yarn or its thickness or even the thickness of textile material in the process earlier to the formation of yarn is governed by many other factors.

It is, therefore, customary to specify the size of yarn (or material at pre-processing stage like lap or sliver or roving) in terms of a number, indicating the mass per unit length or its reciprocal. If it is the former, it is called 'linear density'. In the latter case, the number indicating this is called 'Hank' or 'Count' and is defined by a letter (N).

Accordingly, there are two systems of Yarn Numbering: (1) Direct System (mass per unit length) and (2) Indirect System (length per unit mass). In the case of latter, it was a very popular yarn numbering system with British or continental textile industry. The term 'Hank' is used to define the size of the material in pre-spinning stages (i.e. from blow room lap to speed frame roving). The size of the yarn formed is, however, indicated as the 'Count of' Yarn or Filament. Thus, we specify 'Hank' of lap, sliver or roving whereas we specify 'Count' of carded or combed yarn, double yarn, fancy yarn or man-made filament.

7.2 INDIRECT & DIRECT SYSTEM

7.2.1 INDIRECT SYSTEM

In many of the cotton textile mills, the use of indirect system has been still popular. It is called as 'English Count' system. English count (Ne) is defined as "Number of 840 yards (length) in one pound (lb) weight of material.

Thus, if 'L' is the length of a sample weighing 1 lb. then,

$$Ne = L/840 \qquad (1)$$

If 'L' and 'W' are the length in yards and weight in pounds, respectively, of the material,

$$Ne = \frac{L}{840 \times W} \qquad (2)$$

If the weight is expressed in oz/yard or grains /yard (grs/yd) then,

$$Ne = \frac{16}{840 \times oz/yd} \quad \text{[Note: 16oz = 1lb]} \tag{3}$$

$$OR\ Ne = \frac{7000}{840 \times grains/yd} = \frac{8.33}{grains/yd} \quad \text{[Note: 7000 grains = 1lb]} \tag{4}$$

$$= \frac{8.33}{grains/yd} \tag{5}$$

If the measurements are made in g/m or g/yard or grs/m then,

$$Ne = 0.5902 \div g/m \tag{6}$$

$$Ne = 0.5398 \div g/yd \tag{7}$$

$$Ne = 9.109 \div grs/m \tag{8}$$

7.2.2 OTHER INDIRECT SYSTEMS

Like English system of yarn count, there are other indirect systems used in different textile industries. Whereas, English system is popular in the cotton textile industry, these systems are followed in Woollen and Worsted trades.

$$Nwool = \frac{\text{Length in yards of material}}{256 \times \text{its weight in lb}} \tag{9}$$

$$Nworsted = \frac{\text{Length in yards of material}}{560 \times \text{its weight in lb}} \tag{10}$$

$$Nmetric = \frac{\text{Length in metres}}{1000 \times \text{weight in kg}} \tag{11}$$

$$Ntypp = \frac{\text{Length in yards of material}}{1000 \times \text{its weight in lb}} \tag{12}$$

$$Nfrench\,(Nf) = \frac{\text{Length in metres}}{1000 \times \text{weight in 0.5 kg}} \tag{13}$$

$$= \frac{0.5}{g/m} \tag{14}$$

7.2.3   Important Relations Amongst Different Indirect Systems

$$Nf = 0.847 \, Ne \qquad\qquad (15)$$

$$Nf = 0.5 \, Nm \qquad\qquad (16)$$

7.2.4   Direct System

Unlike indirect system, where the weight of the material is inversely proportional to the yarn number (i.e. higher the number, finer is the material), in Direct System, the weight per unit length of the material is proportional to the yarn number (i.e. higher the number, the coarser is the material). There are mainly two systems internationally followed, viz. 'Tex' and 'Denier'. Thus,

Tex = Weight in grammes of 1000 metres
Denier = Weight in grammes of 9000 metres

7.2.5   Some Other Direct Systems

Grex = Weight in grammes of 10000 metres
Spyndle = Number of pounds per 14,000 yards
Dram = Number of dram weights per 1000 yards [16 dram weights = 1 oz]

7.3   SOLVED EXAMPLES

Example No. 1: Converting a yarn number

Suppose, a yarn of 'T' Tex is to be converted to English count 'Ne' then by definition,

$$Tex = \frac{T\,g}{1000\ m} = \frac{T/453.6\ lb}{1000 \times 1.094\ yards} = \frac{T/453.6\ lb}{1094\ yards}$$

[Note: 1 m = 1.094 yards and 1 lb = 453.6 g]

This means that 1094 yards weigh T/453.6 lbs. Applying the definition of English count, we have to find number of 840 yards in lb. Hence,

$$1lb \text{ of yarn will measure} = \frac{1094 \times 453.6}{T} \text{ yards} \qquad (A)$$

According to definition, therefore, number of 840 yard in (A) will give count (Ne)

$$\text{Number of 840 in 1 lb(Ne)} = \frac{1094 \times 453.6}{T \times 840} = \frac{590.7}{T}$$

Instead of 590.7, the constant which is taken for the use is approximately 590

Thus we have, Ne = 590/Tex

Example No. 2: To derive the equation in a similar manner for Ne, Nf

If one metre length of lap weighs 300 g. Find the French count and English count.

One metre lap weighs 300 g, hence,

500 g of lap would measure = 500/300 × 1 = 1.66 m

By definition, Nf = 1.66/1000 = 0.00166 (French count)

Also, 300 g/m = [300 × 15.43] ÷ [1.094] grs / yd (**Note** 1 g = 15.43 grs)

$$= 4231.26 \text{ grs/yd}$$

By using earlier expression (5), we can get English count.

$$Ne = 8.33 \div \text{grs/yd} = 8.33/4231.26 = 0.00196$$
$$[\textbf{Note: } Nf = 0.847 \times Ne]$$

Example No. 3: A wrapping of six yards of card sliver gave the following readings –330 grs, 348 grs, 312 grs, 324 grs and 330 grs. What is the hank of the sliver?

The average weight of the wrapping is [330 + 348 + 312 + 324 + 330] / 6 = 55 grs/yd

By definition, 55 grs/yd means (7000)/55 yards in 1 lb.

Therefore, number of 840 yds in 1 lb will be: (7000) ÷ (55 × 840) = 0.151 Ne.

The same answer can also be obtained by using the constant 8.33. Thus,

$$Ne = 8.33/55 = 0.151$$

Example No. 4: A lea of ring yarn weighs 15 grains. What is the yarn count in Ne and Tex?

Length of 120 yards is called 'Lea". Similarly, length of 840 yards is called "Hank". It should be noted that this term 'Hank' is different from the term 'hank' (referring to count) used earlier. In fact, the meaning becomes clear from context, e.g. when it is stated as hank of sliver (or lap or roving), it means length-weight relationship to describe fineness of the material (i.e. count). However when it is stated as 'so many hanks produced in a certain time interval' it means merely the length in terms of 840 yards.

Hence, 15 grs/lea means, it is the weight of 120 yards of yarn. Therefore, the length of sliver for 1 lb of yarn will be [7000 ÷ 15] x 120 = 56000 yards of sliver.

As per definition, the number of 840 yards in this length will be =

[56000/840] = 66.66^s which is the count of this yarn in Ne.

By using another short-cut, Ne = [(1000) ÷ grains per lea], we can get the same answer.

Tex of the same yarn = (590/Ne) = 590/66.66 = 8.85

Example No. 5: The hank of roving is 0.8, find its weight in grains for a length of 15 yards.

By definition, 0.8 hank means number of 840 yards in 1 lb

Hence, (840 x 0.8) = 672 yards would weigh 1 lb or 7000 grains.

Therefore weight of 15 yards = [15] ÷ 672] x 7000 = 156.25 grains.

7.4 EXERCISES

1. If 30 m of roving weighs 7 g, what is its hank in French system? (**Ans: 2.14 Nf**).
2. 5 m of drawing sliver weighs 17 g, what is the Nf and Ne? (**Ans: 0.147 Nf; 0.173 Ne**).
3. In the test, 1 yard pieces of blow room laps were weighed. The readings are – 12.2 oz, 12.3 oz, 11.8 oz, 11.9 oz and 12.3 oz. Find the hank of lap. (**Ans: 0.00157 Ne**).
4. One km of yarn weighs 50 g, what are Tex and denier? Also, find the equivalent count in worsted. (**Ans: 50 Tex, 450 den and 17.71 worsted**).
5. A lea of yarn weighs 12.2 grains, what is the English count? (**Ans: 81.96 s Ne**).
6. 40 warp threads, each of 0.1 yard were cut from a fabric sample. The nominal count of warp is 30 s, what will be the weight of these threads? (**Ans: 1.11 grs or 71.93 mg**).

7. If the hank of blow room lap is 0.0012 Ne, find the weight of 40 m length of a full lap. (**Ans: 19.67 kg**).

8. Derive an expression for converting worsted count into English count.

7.5 VARIOUS TYPES OF DRIVES USED[5]

There are various modes of transmissions of power used in textiles machinery when the drive from one part of the machine is conveyed to the other part. The following methods are used depending upon the factors such as – distance between the two parts, direction of the motion, whether motion is continuous or intermittent, reciprocating or variable, whether motion is transferred at right angle or whether there is a drastic reduction in the speeds. Accordingly, the type of drive is chosen.

7.5.1 PULLEY DRIVE

The pulley can be flat or round-grooved or V-shape grooved (Fig. 7.1). Accordingly, they are driven by flat belt, rope and 'V' belt, respectively. In very old machines, rope drive was used, especially when the distances were very large (drive from the overhead shafts) and when the path of the rope was not coplanar or complex. For a long time, individual drives have been steadily introduced in driving the textile machines.

With ropes, the contact between the pulley and rope was less, and hence lot of power was wasted due to slippage.

With leather (flat) belt, the hold on the pulley is improved. Even then, there is a small percentage of slippage and stretching with flat belt, especially the latter due to the change in the condition of the belt (dryness). The width of the flat belt is another criterion that is governed by the power to be transmitted. With the larger power transfers, the width has to be bigger.

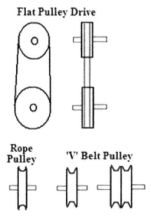

Flat Pulley Drive

Rope Pulley   **'V' Belt Pulley**

FIGURE 7.1 Pulley Drive[1]: A very common & popular way of transferring power over a comparatively longer distance. However, there is power loss due to belt/rope slippage.

The improvement over leather belt is the use of nylon belt on flat pulleys. This belt has much higher strength, reduced slipping tendency, better grip, higher abrasion resistance, less extensibility and, therefore, its life is longer. However, the cost is comparatively higher. 'V'-belts are used for more firm grip and higher load transmission. However, the distance between power transmissions needs to be shorter. V-belts are not used as single belt. Usually, there are two or even three belts used depending upon the power to be transmitted. The grooved pulley accordingly has two or three grooves.

7.5.2 Gear Drive

7.5.2.1 Ordinary and Helical Gears

There are two types of gears, one having ordinary, straight-cut teeth and the other with helically-cut teeth. In most of the common transmissions of power, ordinary gear wheels (Fig. 7.2-A) are used. The use of helical gears (Fig. 7.2-B), however, ensures an extended contact between the teeth of the gears involved. In this case, the changing of the grip from one tooth to the other is smoother. This avoids any kind of jerky motions, especially during starting or stopping of machine.

The noise level is also considerably reduced. The helical gears are, however, costly. In most of the modern machines, therefore, helical gears are used at critical points.

7.5.2.2 Worm & Worm Wheel

This type of drive is used when the speeds are required to be reduced substantially. Normally a single tooth worm is used, though depending upon the need; double worm can be sometimes used. In this case the speed will be doubled. As shown (Fig. 7.3), with one revolution of worm shaft (A), only one tooth of worm wheel (B) is pushed. Another advantage of this type of drive is that the direction of the drive can be changed at right angle. Thus, for a single worm drive:

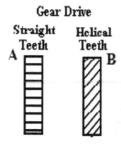

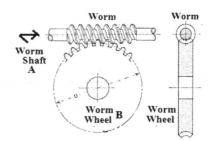

FIGURE 7.2 Gear Wheels[1]: Unlike rope or belt, there is no speed loss during transfer of power. Helical teeth improve the contact between the two gears.

FIGURE 7.3 Worm[3]: It is possible to substantially reduce the speed that is transferred by using this type of drive. The direction of the drive can also be changed.

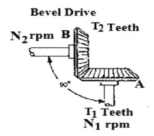

FIGURE 7.4 Bevel Wheels[1]: The teeth of the bevel wheels can either equal or different. But it is possible to change the direction of transfer at right angle.

Speed of Worm Wheel = Speed of Worm (shaft)/Number of teeth of Worm Wheel

7.5.2.3   Bevel Drive

In this case, the shape of the gear wheels is beveled on either side. (Fig. 7.4). The main idea of using this type of drive is to transmit the drive in perpendicular direction. However, the distance over which the power is transmitted is comparatively much shorter. In this case, the slanting of teeth of both the bevels must be such that they mesh with each other perfectly.

$$\text{Speed of B}(N_2) = \text{Speed of A}(N_1) \times [T_1/T_2]$$

7.5.2.4   Chain & Sprocket Wheels and Pawl & Ratchet Drive

In chain drive (Fig. 7.5), as the chain sprockets are used, there is no slippage. However, the transmission at higher loads is not desirable. After a long and continuous use, the chain slackens. This requires removing some of the links of the chain to tighten it. The chain drives are not used when the power transmission is over a long distance or when the speeds are high. This is because one side of the chain has a tendency to often jump and this may put the chain out of sprocket teeth.

$$\text{Speed of B} = \text{Speed of A} \times [T_1/T_2]$$

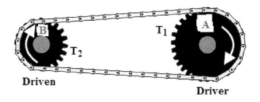

FIGURE 7.5 Chain & Sprocket Wheel[3]: Similar to gear teeth, the chain has also teeth. But the two chain wheels are connected through a chain & can be set at a longer distance.

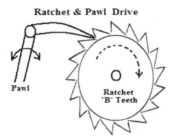

FIGURE 7.6 Pawl & Ratchet[1]: This type of drive initiates a very slow & intermittent motion.

In ratchet and pawl driving (Fig. 7.6), a continuous reciprocating drive is transferred to an intermittent drive. Along with this transformation, there is substantial reduction of the speed as well.

Speed of Ratchet Wheel = Strokes per min of A/Teeth of B

7.5.3 VARIABLE DRIVE

The cone drums are the simple and best example of this type of drive. They are used for imparting variable speed drive (Fig. 7.7). The belt is shifted on the cone drums. This changes the ratio of the driver diameter to driven diameter. Thus, the speed of the driven cone drum can be altered continuously or at intervals and also steplessly. When there is substantially higher load on the cone drum, the belt slips. This is because; the flat belt which is used to drive the cones has to run on tapered surface.

The belt with wider width can be stronger; however, a wider belt reduces the sensitivity in bringing about the necessary changes in the speed. The relative humidity has also its effects in changing the pliability of the leather belt, especially in the low-humid atmosphere, the belt becomes dry. This further leads to more belt slippage. Imperfect alignment of the cone drums also increases load on the belt. The cone drums are required to be maintained, cleaned and their bearings must be periodically oiled for ensuring their smooth running.

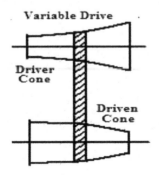

FIGURE 7.7 Cone Drums[1]: It imparts a variable drive, e.g. when a feed is to be varied, the drive is reached through cone drums.

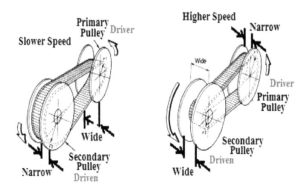

FIGURE 7.8 P.I.V. Drive[3]: In this type of drive, the speeds can be steplessly varied by automatically adjusting the diameters of primary & secondary pulleys. The pulleys have striations on their inside surface and power belt is used to drive them.

In P.I.V. drive used in many modern machines, a similar concept is used. The principle involved is as shown in (Fig. 7.8) where the two pulleys Primary and Secondary are provided with striations on their inclined surface.

These striations on either side, act as gripping surfaces for the belt. For this, therefore, a power grip belt-chain of special type is necessary. There is an arrangement of automatically altering distance between the two flanges of the pulleys and it is synchronised. Thus, the flanges of both the driver and driven pulleys can be moved closer or wider. This changes their working diameter for the belt on the driver and driven pulleys. Accordingly, the speed of the driven changes as follows:

$$\text{Speed of the Secondary (Driven) Pulley} = \frac{\text{Speed of Primary Pulley} \times \text{Its diameter}}{\text{Diameter of Secondary Pulley}}$$

Similar to what is applicable with cone drums, in PIV drive also the condition must be maintained so that $D_1 + D_2 = \text{Constant}$. D_1 and D_2 are correspondingly the diameters of driver and driven pulleys at any working position. The most modern method is to use inverter-motors, where the speeds can be infinitely and very precisely changed.

7.6   BEATS PER INCH OF BLADED BEATER[1]

The capacity of the beater, especially the bladed beater in giving the beating action to the incoming cotton, as mentioned earlier, depends upon its speed, the number and the type of striker blades used and the delivery rate of feed roller. Any of these can be changed so as to vary the intensity of its action. The beats per inch is, therefore, the concept built-up for comparing the actions of the two beaters. Thus,

$$\text{Beats/inch} = [\text{Beats/min}] \div [\text{inches/min}]$$

7.6.1 SOLVED EXAMPLES

Example No. 1: Two bladed beater running at 880 r.p.m. is fed by a feed roller of 2–5/8 inches diameter. The roller shaft has 30[T] worm wheel driven by a double worm that runs at 150 r.p.m. Find the beats per inch.

> Beats per min = Beater rpm × No. of bladed = 880 × 2 = 1760
> Front Roller rpm = [150 × 2]/30 = 10 rpm
> Front Roller Delivery = π × 10 × 2 − 5/8″ = 82.5 inches per min.
> Beats per inch = 1760/82.5 = 21.33

Example No. 2: On two different scutcher lines, there are two-bladed and three-bladed beaters working separately. Their speeds are 1200 r.p.m. and 1050 r.p.m., respectively. Correspondingly, the feed roller deliveries are 60 inches/min and 68 inches/min. Compare effectiveness of their beating action.

> Beats/inch of two bladed beater = [2 × 1200]/60 = 40
> Beats/inch of three blade beater = [3 × 1050]/68 = 46.3

As beats per inch of three-bladed beater are more than that of two-bladed beater, it is expected that its cleaning action will be slightly better than two-bladed beater.

7.7 SENSITIVITY OF FEED REGULATING MECHANISM[1]

This mechanism on scutcher maintains the uniform feeding rate (quantity per unit time) to the beater. As explained earlier, the weight of the cotton fed to beater per unit time is thus maintained at the desired rate. The regulating mechanism is expected to compensate for any variation across and along the fleece length fed to the beater. From time to time, it is necessary to check whether the regulating mechanism works satisfactorily. A simple method used by the technical supervisory staff in the mills is as follows:

Initially, the average weight of the lap being produced is found by observing about 5–8 laps (say 'a' kg). Some material is then taken out during building of a lap from the feeding lattice and is weighed (say 'b' kg). The full lap which is then made is finally weighed (say 'c' kg).

With 100% compensation, the lap weight should have been 'a' again. But there is a shortage which is not compensated = (a − c) kg. This means that the regulating mechanism has been successful in compensating only [b − (a − c)] kg (1).

Expressing (1) as a percentage of weight removed, we come to know how sensitive is the feed regulating mechanism.

$$\text{Sensitivity}\% = \frac{[b-(a-c)]}{b} \times 100$$

7.7.1 Solved Examples

Example No. 1: 900 g of cotton was taken out from the lap fleece being fed to the feed roller of Kirschner Beater. The average weight of the lap just before carrying out the test was 16.5 kg while that made at the end of the test was 16.1 kg. Find the sensitivity of mechanism.

$$\text{Weight not compensated} = 16.5 - 16.1 = 0.4 \text{ kg.}$$

The feed regulating mechanism has, thus, made up for 0.5 kg (0.9–0.4)

$$\text{Sensitivity\%} = (0.5/0.9) \times 100 = 55.55\%$$

Note: In the above example, the sensitivity of 55.55% is considered to be very poor. The satisfactory level is above 88–90%; whereas, a value of 94% and above can be considered to be really good.

Example No. 2: With an average weight of the lap being 18.5 kg, 800 g of material was taken from the feed lattice. What is the expected weight of this lap if expected sensitivity is 85%?

$$[(\text{Weight compensated by mechanism})/(\text{weight taken out})] \times 100 = 85$$

Thus, [(Weight compensated)/800] x 100 = 85

Therefore, the weight compensated = 680 g and that not compensated = 120 g or 0.12 kg

$$\text{Expected weight of the lap} = (\text{Average lap weight}) - (\text{weight not compensated})$$
$$= 18.5 - 0.12 = 18.38 \text{ kg}$$

7.8 DRAFT CALCULATIONS[1,5]

As the material passes through the nip of the rollers with step-up speeds, it gets pulled or attenuated. This motion of drawing or pulling of the stock continues throughout the spinning process. In some cases, there is purposeful attempt made to attenuate the material; while in some others, a very slight pulling action is exercised. At most of the places in the blow room, the latter action is followed, whereas, the former action is used in many subsequent processes till the final yarn is formed. There are three types of drafts:

1. Mechanical Draft (M.D.)
2. Actual or Resultant Draft (A.D./R.D.)
3. Tension Draft

The relative speed difference in the rate of feeding and rate of delivery results in mechanical drafting action. It is governed by the mechanical gears or pulleys.

Mechanical Draft = (Length Delivered) ÷ (Length Fed)
= (Rate of Delivery) ÷ (Rate of Feeding)
= (Surface Speed of Delivery Roller) ÷ (Surface Speed of Feed Roller)

There is no intended mechanical pulling action in the blow room. However, the process, along with carding or combing, extracts process waste, which further reduces the size (weight) of the material delivered. Thus, the resulting hank of the material becomes finer.

$$\text{Actual or Resultant Draft} = \frac{\text{Hank delivered}}{\text{Hank fed}} \times \text{Number of doublings}$$

$$= \frac{\text{Weight fed}}{\text{Weight delivered}} \times \text{Number of doublings}$$

The relation between the Actual Draft and Mechanical Draft is given as follows:

$$\text{Actual Draft} = \frac{\text{Mechanical Draft}}{(100-\text{waste\%})} \times 100$$

It is evident that whenever the process extracts waste, the actual draft will be greater than mechanical draft. However, in speed frame and ring frame, the material receives twist, which results in the contraction along the length. Therefore, the material becomes coarser and in such cases, the **Actual Draft** would become less than mechanical draft.

At many places in spinning, only a slight stretching action is required to be exercised on the material rather than the regular thinning out operation. Such tensioning is necessary to avoid sagging or wrinkling of the material in the process. This controlled stretching is called **Tension Draft**. Whenever mechanical or tension draft is required to be varied, the appropriate change wheels or pinions can be conveniently changed. A common relation which helps in bringing out these changes is as follows:

$$\text{Draft} = \frac{\text{Draft Constant}}{\text{Change Wheel Teeth}}$$

Thus, a draft constant represents a combined cumulative effect of all the gearings and dimensions of the rollers involved in driving the delivery and feed roller. It thus takes into account all the gears in reaching the drive from delivery roller to feed roller, except the change wheel itself.

7.8.1 SOLVED EXAMPLES

Example No. 1: The actual draft in the finisher scutcher is 3.8, while the weight of the lap sheet fed is 12.8 oz per yard and 4 such breaker laps are fed to finisher. If the process removes 2.1% waste, calculate the mechanical draft and weight per yard of the lap delivered.

$$\text{Mechanical Draft} = [\text{Actual Draft} \times (100 - \text{waste}\%)] \div 100$$
$$= [3.8 \times (100-2.1)] \div 100 = 3.72$$

Also, Actual Draft = [weight fed/yd ÷ weight delivered/yd] × No. of Doublings
Hence, 3.8 = (12.8 × 4)/weight delivered
Weight delivered = (12.8 × 4) / 3.8 = 13.47 oz/yd

Example No. 2: The final lap dimensions are 38-inch width and 26-inch diameter. If the full lap weight is 40 lb, find the density of lap.

$$\text{Volume of lap} = \pi \times [\text{diameter}/2]^2 \times \text{width}$$

$$\text{Volume of lap} = \pi \times [26 \div (2 \times 36)]^2 \times 38/36 \text{ cubic yard}$$
$$= \pi \times 0.13 \times 1.005 = 0.4096 \text{ cubic yard}$$

Therefore, Lap Density = Mass/Volume = 40/0.4096 = 97.65 lb/cubic yard

Example No. 3: A blow room lap of 300 g/m is delivered by a calendar roller of scutcher. The waste extracted at three-bladed beater and Kirschner beater taken together amounts to 1.8%. The mechanical draft between the feed roller of 3 B.B. and cage delivery roller of Kirschner Beater is 2.7 while that between cage delivery roller of Kirschner Beater and bottom Calender Roller is 1.4. Find the actual draft. Also, find the weight of the material on feed lattice of Three Bladed Beater.

$$\text{Total Mechanical Draft} = 2.7 \times 1.4 = 3.78$$

$$\text{Actual Draft} = (\text{Mechanical Draft} \times 100)/(100 - \text{waste}\%)$$
$$= (3.78 \times 100)/(100 - 1.8) = 3.849$$

Weight of the material on the lattice of Three Bladed Beater will be:

Weight of lap delivered by calendar roller x Actual Draft

$$= 300 \times 3.849 = 1154.7 \text{g/m}$$

7.8.2 EXERCISES

Exercise No. 1: The weight of the lap delivered by a scutcher is 13.8 oz/yard. The draft between C.R. and feed roller feeding to the beater is 4.2. The tension draft between the lap roller and C.R. is 1.1 and waste removed by the beater is 2%. Find the actual draft and weight/yard of the lap sheet as delivered by the feed roller.

(Ans: Act, Draft = 4.7; Weight = 64.99 oz/yard).

Exercise No. 2: Four laps each of 14.5 oz/yard are fed to finisher scutcher. If the waste extracted is 1.7% and the lap delivered is 15 oz/yard, find the actual and mechanical draft.

(Ans: Act. Draft = 3.86; Mech. Draft = 3.79)

Exercise No. 3: In a finisher scutcher, the weight of the material fed at feed roller is 1250 g/m. If the mechanical draft is 4 and the weight per unit length of the lap delivered by calendar roller is 300 g/m find the actual draft and waste extracted.

(Ans: Actual Draft = 4.16; waste = 4.1%)

7.9 TYPICAL MOTIONS IN CONVENTIONAL BLOW ROOM[1]

7.9.1 LOAD ON CALENDER ROLLERS

As shown in Fig. 7.9, the lap enters the system at point between top and second calendar roller. Finally, curving around the second and third C.R., it finally comes out through third and bottom C.R.

Thus, the lap at A has the pressure of three calendar rollers. In addition to this, the weighting through the side lever on either side puts additional pressure.

The total pressure on the lap, therefore, can be easily calculated, taking the moment of weight W_1:

Downward pull at K = 2 ($W_1 \times$ BF / KF) – (say M)

Total Load on the Lap = 3 × W_2 + M

[**Note:** Here the weight of each calendar roller is assumed to be W_2; also the weight W_1 is on either side of the machine.]

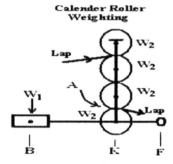

FIGURE 7.9 Pressure on Lap[1]: This makes the lap more compact helps to reduce lap licking tendencies.

Example No. 1: Calculate the total load on the lap when each calendar roller weighs 130 lb. The weight of 25 lb is hung at a distance 4.8 feet (BF – Fig. 7.9) from the fulcrum of weight-lever. The link connecting the C.R. and weight-lever is attached to the latter at a distance 1½ inch (KF in Fig. 7.9) from the fulcrum.

Load of three C.R. = 3 × 130 = 390 lb
Load due to weighting system = (25 × 4.8) ÷ (1.5/12) = 960 lb

As the weighting system is on either side of calendar roller, the weighting due to hanging weight on lever will be doubled

Total load = 390 + 2 × 960 = 2310 lb

7.9.2 LAP LENGTH MEASURING MOTION[1]

It measures a pre-determined length of the lap delivered by calendar rollers and immediately brings about either stoppage of the machine (as in old conventional scutcher) or when auto-doffing device is in operation, the machine without stopping automatically doffs the lap, replaces the lap spindle and starts a new lap.

7.9.2.1 Mechanical Type

The knock-off wheel (Fig. 7.10) plays an important role. After every one complete rotation of this wheel, the knock-off is brought about, involving the displacing of the starting handle, which stops the lap rollers (old scutcher).

One revolution of this wheel means

(K / C) x (W / teeth on worm) = P (Number of revolutions of bottom calendar roller [**Note: C = change pinion**]

Hence, the length of the lap delivered by calendar roller = (π × dia. of C.R. × P) / 36 **yards**.

Replacing value of 'P' by assuming a single worm

$$= [\pi \times \text{dia. of C.R} \times K \times W] / 36 \times C \qquad (1)$$

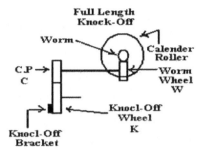

FIGURE 7.10 Lap Length Motion[1]: Precise length of lap makes every lap identical. Thus, with every lap weighed, it indicates variation in its weight per unit length.

In the expression (1), the numerator, including number 36 is constant for a set of wheels used by the manufacturer, and is called "Lap Length Constant". The wheel C is a change pinion and is changed to alter the length of the lap for each knock-off. The expression, therefore, can be summarised as:

Length of Lap = Lap Length Constant/Change Pinion

7.9.2.2   Hunter & Cog Knock-off Motion[5]

In yet another mechanism, the bottom calendar roller carries two compound wheels, one of which is a **cog** wheel (Fig. 7.11a) and the other is gear wheel A (Fig. 7.11b) behind this cog wheel. Wheel A drives yet another gear B, which is compounded with hunter wheel.

After every T_1 revolutions of gear B or T_2 revolutions of gear A, the hunter and the cog face each other. This results in cog pushing the hunter away. The hunter wheel conveys this pushing motion through a link to the starting handle which ultimately gets knocked-off.

Thus, the length of the lap per knock-off will be

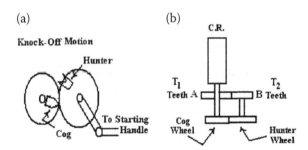

FIGURE 7.11 Cog & Hunter Full Length Stop Motion 1: A combination of Odd & Even number of teeth on the two wheels bring cog & hunter opposite each other after the delivery of precise length of lap. This leads to mechanical knock-off (putting off the starting handle): (a) Cog & Hunter; (b) Cog & Hunter gearing.

$$\text{Lap Length} = \frac{(\pi \times \text{dia. of C.R.}}{36} \times \frac{T_1 \times T_2}{T_2} = [\text{Lap Length Constant}] \times T1$$

7.9.3 SOLVED EXAMPLES

Example No. 1: In Hunter & Cog mechanism, the gear wheel on C.R. (cog wheel) has 50 teeth. The lap length constant is 0.9 whereas the diameter of C.R. is 5 inches. Find the length and weight of this lap, if its weight per yard is 13.5 oz.

Lap Length = Constant x Teeth on Cog Wheel

= 0.9 × 50 = 45 yards

Weight of the Lap = (45 ×13.5) ÷ 16 = 37.96 lbs or 17.21 kg

Note: For bringing the knock-off, only the teeth of the cog wheel carried by calendar roller are important in deciding the length of the lap. By varying the teeth on this wheel, both the length and hence the lap weight can be changed. The value of the other wheel is usually a few **teeth** higher or lower than cog wheel; or else, the combination of the wheels should be such that their L.C.M. is the product of number of teeth on both. In short, the L.C.M. for the wheels shown in the (fig. 144/145) = T_1 x T_2 where either T_1 or T_2 is a **prime number**.

Example No. 2: In Hunter & Cog mechanism, the cog wheel on C.R. has 61^T. If the length of the lap is 40 yards and the hunter wheel working in conjunction with the above cogwheel has 50^T, find the diameter of C.R.

After every 61 revolutions of hunter wheel, the knock-off would take place. Hence for every knock-off

The diameter of the calender roller can be calculated as follows:

$$\text{Length of Lap} = \frac{\pi \times \text{dia. of C. R.} \times 61}{36} = 40 \text{ yards}$$

Hence,

$$\text{Diameter of C.R.} = (40 \times 36) \div (\pi \times 61) = 7.5 \text{ inch}$$

Example No. 3: A double worm on calendar roller drives a worm wheel (Fig. 7.10) of 64^T. The shaft carrying the latter carries a change pinion of 27^T

which gears with the knock-off wheel of 48^T. With diameter of C.R. as 7 inches, find the length of the lap delivered after every knock-off.

For one revolution of knock-off wheel

$$\text{No. of Revolutions of C.R.} = \frac{48 \times 64}{27 \times 2} = 56.88$$

Thus, the length of lap delivered by C.R. = $[56.88 \times 7 \times \pi] \div 36 = 34.75$ yard

$$\text{Lap Length Constant} = \text{Length of Lap} \times \text{Change Pinion}$$
$$= 34.75 \times 27 = 938.25$$

7.9.4 EXERCISES

Exercise No. 1: The lap length, with worm & worm wheel measuring motion, is 41 yards. The teeth on the knock-off wheel & worm wheel are 60^T & 36^T, respectively. There is a single worm on calender roller which has 5 inches diameter. Find the lap length constant and find the change pinion. **[Ans: 942.6; 22.99 or 23^T]**

Exercise No. 2: Calender roller of 7″ diameter has a single worm driving 27^T worm wheel. The change wheel of 21^T on this shaft drives 49^T knock-off wheel. If the draft between the calendar roller and lap roller is 1.05 and the hank of the lap is 0.00128, find the total lap length and its weight. **[Ans: 40.4 yards; 37.57 lb]**

Exercise No. 3: In Hunter & Cog measuring motion, the cog wheel on C.R. has 79^T; whereas the hunter wheel has 80^T. The stretch in the lap between lap roller and C.R. is 2.5%. If the hank of lap is 0.00133 and dia. of C.R. is 6.8″, find the length of lap and its weight. **[Ans: 48.05 yards & 43.28 lb]**

Exercise No. 4: The lap length constant for Hunter and Cog measuring motion is 0.560, what is the value of cog wheel on C.R. to give 40 lb weight of the total lap with its hank as 0.0012? **[Ans: 72^T]**

7.10 SCUTCHER PRODUCTION[1]

The production of scutcher can be calculated from the speed of either calender roller or lap roller. The lap length delivered by lap roller per minute will be

$$= \frac{\pi \times \text{L.R. rpm.} \times \text{its dia.}}{36} \quad \text{where as, weight of the lap will be} -$$

$$\text{Weight delivered/min} = \frac{\pi \times \text{L.R. rpm} \times \text{its dia.} \times \text{Wt. /yard}}{36} = (\text{say} - -\text{B})$$

If the working efficiency is $\eta\%$, and the running time is a shift of 8 hr, the production of the scutcher will be

$$\text{Production/shift} = \frac{(\text{B}) \times \eta \times 60 \times 8}{100} \qquad (1)$$

Depending upon the unit of weight per unit length (oz/yd or g/m), the value of the expression (1) can be expressed in either pounds per shift or kg per shift by using appropriate conversion factor.

However, when the production is calculated from C.R., a correction factor equal to the draft between calender roller and lap roller must be applied.

Actual Production = Production from C.R. × Draft between L.R. & C.R.

7.10.1 Solved Examples

Example No. 1: Find the speed of the calender roller having 6.5″ diameter when a lap of 44 yards is produced in 4.5 min and the draft between Lap Roller and Calender Roller is 1.03.

Surface Speed of L.R. = 44 yards per 4.5 minutes
= 9.777 yd/min

Surface of C.R. as calculated from Lap Roller, taking into consideration the draft in between will be – 9.777 / 1.03. Therefore,

$$\frac{9.777}{1.03} = \frac{\pi \times \text{C.R.rpm} \times \text{its dia.}}{36} = \frac{\pi \times \text{C.R.rpm} \times 6.5}{36}$$

Hence, C.R. r.p.m. = 16.73

Example No. 2: Find the production of a scutcher per shift of 8 hr, if 10″ diameter of lap roller, running at 9 r.p.m. delivers a lap of 14.2 oz/yd. The working efficiency of the machine is 90%.

$$\text{Surface speed of L.R.} = \pi \times \text{L.R. dia.} \times \text{its rpm}$$
$$= (\pi \times 10 \times 9)/36 = 7.855 \text{ yd/min}$$

$$\text{Prod. of Scutcher} = \frac{\text{Surface Speed of L.R.} \times 60 \times 8 \times 14.2 \times 90}{16 \times 100} = 3011.6 \text{ lb/shift}$$

OR = 1365.8 kg/shift

Example No. 3: A scutcher producing 1200 kg/shift and working with 88% efficiency delivers a lap of 0.00125 hank. If the calender roller has 7″ dia. and the draft between Lap Roller and Calender Roller is 1.07, find r.p.m. of Calender Roller.

$$\text{Production} = \frac{\text{S.S. of Lap Roller} \times 60 \times 8 \times \text{efficiency}}{36 \times 840 \times \text{hank of lap} \times 2.205}$$

$$1200 = \frac{\text{S.S. of Lap Roller} \times 60 \times 8 \times 88}{36 \times 840 \times 0.00125 \times 100 \times 2.205}$$

Hence,

$$\text{Surface Speed of Lap Roller} = 237.73 \text{ inch/min}$$

Taking into consideration the draft of 1.07 in between Lap Roller and Calender Roller, the surface speed (S.S.) of Calender Roller will be

$$\text{S.S. of C.R.} = \frac{\text{S.S. of Lap Roller}}{\text{Draft}} = \frac{237.73}{1.07} = 207.64 \text{ inch/min}$$

$$\text{C.R. rpm} = \frac{\text{S.S. of C. R.}}{\pi \times \text{C.R. dia.}} = \frac{207.64}{\pi \times 7} = 9.44 \text{ rpm}$$

7.10.2 EXERCISES

Ex. No. 1: A scutcher has a lap roller of 11″ diameter and it runs at 10 r.p.m. The hank of the lap produced is 0.0013. Find the production per shift at 80% efficiency **[Ans: 1531.07 kg]**

Ex. No. 2: A calender roller of scutcher runs at 15 r.p.m. and has 7″ diameter. The draft between lap roller and calender roller is 1.04. If the lap weight is 13 oz/yd and the total lap weight is 40 lb, find the production and length of each lap. **[Ans: 1685 kg; 49.23 yards]**

Ex. No. 3: Find the time for completion of one lap when a lap roller of 10.5″ diameter. runs at 11.5 r.p.m. The hank of the lap produced is 0.00123 and its total weight is 35 lb **[Ans: 3.43 minutes]**

Ex. No. 4: A scutcher produces a lap in 4.2 min. If the lap weight is 18 kg and lap roller diameter is 25 cm, find the r.p.m. of lap roller and the production per shift at 82% efficiency, when hank of lap is 0.0014 **[Ans: 12.93 r.p.m.; 1686.85 kg]**

Ex. No. 5: A lap roller of 24 cm diameter and running at 14 r.p.m. produces a lap of 300 ktex, find the production per shift. **[Ans: 1520.22 kg]**

7.11 WASTE EXTRACTION[1,5]

As explained in the earlier chapter, the beaters in the blow room extract waste. The usual method of finding out the percentage of trash (droppings) extracted under the beater is to feed a certain quantity of cotton (say 'Y' kg) to the beater and then collect the droppings (say 'X' kg) under it, at the end of the test.

The trash extracted by the beater is then expressed as Waste% = (X × 100) / Y

In a blow room, the cleaning efficiency of the machine depends upon the total trash extracted by the various beaters employed in a sequence. However, trash extracted by a single machine depends upon several other factors such as (1) The nature and construction of the openers/beater incorporated (2) Its position in the sequence (3) The dust separating contrivance associated with the machine (4) The trash in the material fed (5) The speed and settings of grid bars with respect to beater and (6) The speed with which material travels through the beater. If the cotton fed to the first machine is (say) 'Y' kg and the droppings under the various beaters in the sequence are – (say) X_1, X_2, X_3, then, the percent droppings for each beater will be

$$1^{st} \text{ Beater} = \frac{X_1 \times 100}{Y}; \quad 2^{nd} \text{ Beater} = \frac{X_2 \times 100}{Y-X_1}; \quad 3^{rd} \text{ Beater} = \frac{X_3 \times 100}{Y-(X_1 + X_2)}$$

[Note: Here invisible loss is not taken into consideration]

7.11.1 SOLVED EXAMPLES

Example No. 1: In a certain study, 10 bales of 180 kg each of 20 s mixing were processed through a blow room line. The grid bars and the trash chamber under the beaters were initially cleaned. After processing 10 bales, the droppings under each beater were collected. The data for the collected droppings is as follows:

Hopper Bale Breaker – 11.5 kg; Step Cleaner – 24.8 kg; Crighten Opener – 36.7 kg

Porcupine Opener – 33.9 kg; Three Bladed Beater – 12.8 kg and Kirschner Beater – 12.9 kg. Find the total and individual waste%.

TABLE 7.1
Beater-Wise Waste %

No.	Machine	Material Fed	Waste Taken-Out	Waste%
1.	Hopper Bale Breaker	1800.0 kg	11.5 kg	0.63
2.	Step Cleaner	1788.5 kg	24.8 kg	1.38
3.	Crighten Opener	1763.7 kg	36.7 kg	2.08
4.	Porcupine	1727.0 kg	33.9 kg	1.96
	Scutcher No.1	Approx. half of above fed to Scutcher No.1		
5a.	Three Blade Beater	863.50 kg	12.8 kg	1.48
6a.	Kirschner	850.70 kg	12.9 kg	1.51
	Scutcher No.2	The remaining half fed to Scutcher No.2		
5b.	Three Blade Beater	863.50 kg	12.8 kg	1.48
6b.	Kirschner	850.70 kg	12.9 kg	1.51

Note: A blow room line sequence invariably bifurcates the feed material prior to scutcher. Each scutcher line is thus expected to get half the share. As in the above example, there are two scutcher-lines, the weight fed to each is halved.

$$\text{Total Waste Extracted} = 11.5 + 24.8 + 36.7 + 12.8 + 12.9 + 12.8 + +12.9$$
$$= 124.4 \text{kg}$$

$$\text{Total Waste\%} = [(124.4)/1800] \times 100 = 6.91$$

Further, had the waste percentage at every beater been calculated in the above example by taking the material fed as 1800 kg for each machine, the value of the waste% for each beater would have varied only marginally.

For Example,

$$\text{Waste \% at Crighten} = \frac{\text{Waste extracted at Crighten}}{\text{Total cotton fed}} = \frac{36.7 \times 100}{1800} = 2.03\%$$

It can be easily seen that this value when compared with that given in Table 7.1 (which is 2.08%) is not significantly different.

Example No. 2: The scutcher delivers a lap of 450 g/m. The total draft between lap roller and feed roller of Three Bladed Beater is 4.14. The weight of the material under the above feed roller is 1900 g/m. Find the percent waste removed under both Three Bladed Beater and Kirschner Beater taken together.

$$\text{Actual Draft} = \frac{\text{Weight fed at feed roller}}{\text{Weight delivered at lap roller}} = \frac{1900}{450} = 4.22$$

$$\text{Actual Draft} = \frac{\text{Mechanical Draft}}{(100-\text{Waste}\%)} \times 100$$

Hence,

$$(100-\text{Waste}\%) = \frac{\text{Mechanical Draft}}{\text{Actual Draft}} \times 100 = \frac{4.14 \times 100}{4.22} = 98.1$$

$$\text{Waste}\% = 100 - 98.1 = 1.9\%$$

7.12 CLEANING EFFICIENCY[1,5]

When the beater extracts the trash, it results in cleaning of the stock processed. This can be represented in terms of cleaning efficiency of the machine. Thus, the cleaning efficiency (C.E.) can be expressed as

$$\text{C.E.} = \frac{\text{Trash in material fed} - \text{Trash in the material delivered}}{\text{Trash in the material fed}} \times 100$$

For finding the trash in the material fed to the machine and that delivered by it, these samples are processed through Shirley Analyzer, which separates the lint from trash. The trash so separated is collected and weighed. The weight of the trash can then be expressed as percentage of trash content in the material tested.

7.12.1 SOLVED EXAMPLES

Example No. 1: Two cotton samples each of 100 g and taken from two different bales were processed separately through Shirley Analyzer. The readings are given in Table 7.2. Find the trash content in the cotton.

TABLE 7.2
Lint & Trash Composition

Sample No.	Lint	Trash	Invisible Loss
1	93.8 g	4.6 g	1.6
2	93.9 g	4.4 g	1.7

It is customary to take minimum two readings of Shirley Analyzer for each variety of cotton. Assuming that the two bales represent the same variety of cotton, two readings from two different bales are taken in the above sample. Also, it is necessary to pick-up the representative sample of each bale (small portions are taken from different parts of a bale) when testing the trash content. Therefore, if the two readings of Shirley Analyzer differ significantly, it means that either the sample is not representative of a bale or else, the two bales differ in their trash content. However, when the readings do not differ very much, the mean of the two readings represents the average trash content of the material tested.

In the above example, the mean of the values of trash content 4.6% and 4.4% (not differing much) is 4.5% and represents the trash content of the cotton fed.

After the lint and trash are separately collected in Shirley Analyzer, the sum total of the weights of these two is always less than the material fed (i.e. <100 g). This is because very fine dust along with short fibrous matter is likely to follow the path of the suction generated by a fan within Shirley Analyzer. This is obviously not accounted for. Thus, the difference of 1.6 or 1.7 in the above example in the two readings corresponds to **'Cage Loss'** or **'Invisible Loss'**. With higher trash content in the tested material, the cage loss is proportionately higher. In short, the cage loss is higher for the bale cotton than for the laps. However, a cage loss of more than 2% also would mean that the portion of trash which normally should have fallen down in the tray under the Shirley Analyzer is lost through the cage perforations. In such cases, the readings are discarded and after readjusting the suction, a fresh reading is taken again.

Example No. 2: Samples from bale cotton and laps were tested on Shirley Analyzer. They gave 4.2% and 1.1% trash, respectively. Find the Blow Room cleaning efficiency.

$$\text{Cleaning Efficiency} = \frac{\text{Trash in bale} - \text{Trash in lap}}{\text{Trash in bale}} \times 100$$

$$= [(4.2 - 1.1)/4.2] \times 100 = 73.8\%$$

Example No. 3: Cotton fed to Crighten Opener contains 5.2% trash. The beater extracts 2.5% waste which contains 80% trash. Find the cleaning efficiency.

Out of 2.5% waste, 80% is the trash, Hence,

Actual trash removed = [(2.5 × 80/100] = 2.0%. This is the actual trash removed in Crighten Opener. Therefore, the material delivered will have 3.2% trash.

Cleaning Efficiency = (5.2 − 3.2)/5.2) x 100 = 38.46%

Example No. 4: A Blow Room gives 70% cleaning efficiency, whereas card for processing the same laps gave 80% cleaning efficiency. What is the overall (combined) cleaning efficiency?

Cleaning efficiency of 70% for Blow Room means that 30% trash is still left over in laps. However, the card is able to remove only 80% of this left-over trash. Hence,

Trash extracted by the Card = [(30 × 80)/100)] = 24%

The overall cleaning efficiency of Blow Room and Card taken together will be

70% + 24% = 94%

7.12.2 EXERCISES

Ex. No. 1: Shirley Analyzer test gave trash content as follows: (1) Raw Cotton = 5.7% (2) Blow Room Lap = 1.8% and (3) Card Sliver = 0.4%. Find the Blow Room and Card cleaning efficiency. **[Ans: B.R. = 68.4%; Card = 77.7%]**

Ex. No 2: Two cotton bales were taken for testing the trash content. Both the bales had similar cotton variety. The samples were processed through Shirley Analyzer. The weight of each sample taken for the test was 100 g

Sample No. 1: Lint = 93.2 g; Trash = 4.7 g

Sample No. 2: Lint = 93.8 g; Trash = 4.2 g

Find the average trash content and invisible loss.

[Ans: Avg. Trash Content = 4.45%; Inv. Loss = 2.05%]

Ex. No. 3: Bale cotton containing 8.6% trash was fed to Blow Room. The laps made were processed on Card. The cleaning efficiency levels are Blow Room = 75%, Card = 81%. Find the trash in the lap and sliver. [Lap = 2.15%; Sliver = 0.41%]

7.13   LINT LOSS[2]

The analysis of droppings under the beater often gives interesting information about the proportion of lint and trash. Thus, the loss of lint along with the extraction of trash is unavoidable. Therefore, whenever the efforts are made to increase the cleaning efficiency of individual beater or overall blow room process, the balance of both, the lint and trash in the droppings must be carefully watched and maintained. However, it may be noted that a significant increase in the proportion of trash at a cost of only a marginal rise in the lint in the droppings may be advisable, if economical. The quality and cost of the raw material being processed will govern the whole economics.

Let, X% = trash in bale cotton

Y% = trash in lap

Z% = total blow room droppings

If 100 kg of bale cotton is fed to blow room (Bale Breaker or Blender), then, (100 – Z) kg of cotton is delivered as laps. This cotton contains Y % trash, hence

$$\frac{(100 - Z) \times Y}{100} \text{kg will be the trash in this lap cotton} \tag{1}$$

With X % trash in the bale cotton and trash as expressed in (1) in lap cotton,

Trash in the droppings under the blow room machines will be

$$\left[X - \frac{(100 - Z) \times Y}{100} \right] \text{kg trash in the droppings} \tag{2}$$

As mentioned earlier, a small proportion of this trash which is very fine is sucked through the cages in blow room and therefore, cannot be accounted for (Cage loss or Invisible loss). If (and when) the proportion is very small, it can be neglected. Thus, the expression (2) when defined as a ratio of the total droppings in the blow room would give trash % in the droppings.

$$\frac{X - [(100 - Z) \times Y]/100}{Z} \times 100 = \%\text{trash in the droppings} \tag{A}$$

@ - Process Control in Spinning – ATIRA Publication

$$\text{Hence, } [100 - (A)] = \%\text{lint in the droppings} \tag{B}$$

7.13.1   SOLVED EXAMPLES

Example No. 1: A cotton mixing has 5.2% average trash content. The trash in the lap is 1.8%; whereas the total blow room droppings are 6.7%. Find the lint loss.

$$\text{Trash in the droppings} = \frac{5.2 - [(100{-}6.7) \times 1.8]/100}{6.7} \times 100 = 52.68\%$$

Therefore, lint loss = 100 − 52.68 = 47.32%

Example No. 2: The laps fed to two different cards have 2.2% trash. The card sliver contains 0.38% trash in one case and 0.25% trash in the second case, with total card waste as 4% and 5% in these two cases, respectively. What will be the lint loss in these two cases?

First Case:

$$\text{Trash content in Card waste} = \frac{2.2 - [(100 - 4) \times 0.38]/100}{4.0} \times 100 = 46\%$$

Hence, lint loss = 100 − 46 = 54%

Second Case:

$$\text{Trash content in Card waste} = \frac{2.2 - [(100 - 5) \times 0.25]/100}{5.0} \times 100 = 39.2\%$$

Hence, lint loss = 10 − 39.2 = 60.8%

TABLE 7.3
Higher Card Waste & Trash in Sliver

Process	Total Card Waste %	Trash in Sliver %	Lint Loss %	Trash Loss %
1st Case	4.0	0.38	54.0	46.0
2nd Case	5.0	0.25	60.8	39.2

As seen from the Table 7.3, the extra card waste extracted in second process has gone to reduce the trash content in the sliver delivered. However, the lint proportion in the card waste has also increased. The actual economics of getting a comparatively cleaner sliver at the cost of higher lint loss has to be worked out by considering on the one hand, improvement in the quality of the final yarn and on the other hand, the cost of higher lint loss.

7.13.2 EXERCISES

Ex. No. 1: The bale cotton has an average trash of 6%. The total blow room droppings are 10%; whereas the trash in the lap is 2.3%. Find the lint loss. [Ans: 60.7%]

Ex. No. 2: The card sliver contains 0.2% trash. The lap fed to it contains 1.5%

trash. If the total card waste is 4%, find the lint and trash percentage in the total card droppings. **[Ans: lint – 67.3% & trash – 32.7%]**

Note: The card waste is mainly composed of two types of waste: (a) licker-in dropping which has much higher trash content and (b) flat strip which has comparatively much less trash content.

Ex. No. 3: The trash content in bale cotton, lap and sliver are 7.2%, 1.9% and 0.31%. If the total blow room loss and total carding loss are 9% and 4.8%, respectively, find the lint and trash content in both. **[Ans: B.R. – 38.89% lint and 61.11 trash; Card – 33.54% lint and 66.46% trash]**

Ex. No. 4: The average mixing trash in Blow Room is 5%. The total droppings when increased from 6% to 7% result in reduction of trash in the lap from 1.7% to 1.4%. Find the lint and trash in the droppings in these two cases. **[Ans: (1) With 6% Droppings: 43.17% lint and 56.83% trash and (2) With 7% Droppings: 47.15% lint and 52.86% trash]**

LITERATURE REFERRED

1. Elements of Cotton Spinning –Blow Room – Dr. A. R. Khare.
2. Process Control in Spinning – ATIRA Publication.
3. Wikipedia.
4. Cotton Spinning – William S. Taggart.
5. Spinning Calculations – Dr. H. V. Sreenivasa Murthy & Dr. A. R. Khare.

Index

Printed in the United States
by Baker & Taylor Publisher Services